I0605798

JAPANESE SPY GEAR & SPECIAL WEAPONS

Scientists and Technicians of the Japanese Army's Noborito Research Institute in the Second World War and the Cold War

JAPANESE SPY GEAR & SPECIAL WEAPONS

HOW NOBORITO'S SCIENTISTS AND TECHNICIANS SERVED IN THE SECOND WORLD WAR AND THE COLD WAR

STEPHEN C. MERCADO

Pen & Sword
MILITARY
AN IMPRINT OF PEN & SWORD BOOKS LTD.
YORKSHIRE – PHILADELPHIA

First published in Great Britain in 2025 by
PEN AND SWORD MILITARY
An imprint of
Pen & Sword Books Limited
Yorkshire – Philadelphia

ISBN 978 1 03610 798 7

A CIP catalogue record for this book is available from the British Library.

Typeset in Times New Roman 11.5/14 by
SJmagic DESIGN SERVICES, India.
Printed and bound in the UK by CPI Group (UK) Ltd.

The Publisher's authorised representative in the EU for product safety is Authorised Rep Compliance Ltd., Ground Floor, 71 Lower Baggot Street, Dublin D02 P593, Ireland.
www.arccompliance.com

For a complete list of Pen & Sword titles please contact
PEN & SWORD BOOKS LIMITED
George House, Units 12 & 13, Beevor Street, Off Pontefract Road,
Barnsley, South Yorkshire, S71 1HN, England
E-mail: enquiries@pen-and-sword.co.uk
Website: www.pen-and-sword.co.uk

or

PEN AND SWORD BOOKS
1950 Lawrence Rd, Havertown, PA 19083, USA
E-mail: uspen-and-sword@casematepublishers.com
Website: www.penandswordbooks.com

CONTENTS

PREFACE

> The materials used in clandestine warfare, truly great in their variety, require the skilled use and application of scientific factors. Depending on the situation, one must bring together the best of contemporary science and go to the top. The Noborito Research Institute was the only institute born at that time from the need for the research, invention, and production of the materials necessary for clandestine warfare.
>
> – Major Ban Shigeo, director, Group 1 of Second Section, and Lieutenant General Shinoda Ryo, director, Noborito Research Institute[1]

Spy gear and special weapons are crucial to intelligence and secret operations. A mask disguises an operative's appearance. Secret ink leaves no visible trace of a message hidden between the lines of a seemingly innocent letter. A camera disguised as a cigarette lighter allows an agent to take pictures without attracting attention. Behind enemy lines, commandos wreak havoc with devices to start fires, explosives built to resemble coal, or pens containing bacteria to infect an enemy's water supply. Operatives pass through checkpoints with forged documents and buy supplies with counterfeit currency. Such equipment, weapons, and materials come from special laboratories established to support intelligence gathering and clandestine operations.

Who makes all this? Q (short for Quartermaster, chief of Q Branch, later Q Division) and his technicians develop for the British agent James Bond the devices for his missions in the popular spy movies. But audiences have seen much more of the dapper 007 than the snappish Q, whose main role on screen is to call Bond to attention before demonstrating his latest gadgets.

In fact as in fiction, details on the technical side of spying and clandestine operations have been scarce. Writers have penned a few histories of the Second World War with reference to the Research and Development Branch (R&D) of Washington's Office of Strategic Services (OSS) and Stations 12 and 14 of London's wartime Special Operations Executive (SOE). On the Axis side, there has been some writing on Berlin's laboratory in Tegel for sabotage techniques, which operated under the Military Intelligence Service (Abwehr) of the German Armed Forces, and Abwehr I G, which was responsible for document forgeries and secret inks.[2]

In the case of Japan, little has appeared in print on the Noborito Research Institute, which developed spy gear and special weapons for the Imperial Japanese Army (IJA). Most of the few books on Noborito are of relatively recent date, published after the end of the Cold War. All have been written in Japanese.

The Empire of Japan came under pressure at the end of the First World War to keep pace with the West in the march of military science. Tokyo had emerged victorious from the war as a member of the Allied side but faced the challenge of developing poison gas and other modern weapons to match those that appeared on the battlefields of Europe. The Japanese Army responded in 1919 by establishing the Army Scientific Research Institute (ASRI), with departments for physical and chemical research. In 1925, the institute added a department for chemical warfare.

The Japanese Army also increased its efforts on covert and special operations. In 1927, ASRI started a small laboratory for clandestine warfare. In 1939, the ASRI's laboratory for clandestine warfare moved to the institute's branch in Noborito, an area of Kawasaki, southwest of Tokyo. In 1942, with the ASRI's reorganisation, the branch became the 9th Army Technical Research Institute (ATRI). By war's end, the 9th ATRI, better known as the Noborito Research Institute, would grow to nearly a thousand employees.

Only the 9th ATRI, among the IJA's 10 numbered institutes, operated under the clandestine operations section of Second Bureau in the Army General Staff (AGS). Noborito technicians worked to develop a variety of tools and weapons, from secret inks and concealed cameras to disguised explosives, a death ray, poisons, bombing balloons, and counterfeit currencies. Personnel were working on a death ray when the war ended. Noborito spy gear helped the Kempeitai and the Yama Agency to counter foreign spies and track domestic dissidents. Operatives of the Nakano School took Noborito explosives and toxins to the field. IJA aircraft attacked Chinese crops with biological weapons from Noborito. Poisons from the institute became tools of assassination. Bombing balloons developed at Noborito struck the United States from across the Pacific Ocean. The institute's counterfeiters produced bogus bank notes in a bid to wreck the Chinese economy. The institute developed those devices in consulting top scientists and technicians in Japan's universities and corporations. Noborito, enjoying the largest budget of the Army's 10 ATRIs, had the means to hire top talent on contract.

When Tokyo surrendered in 1945 and came under Allied occupation, American officials questioned Noborito officers and technicians to learn their secrets. Washington then enlisted some of the institute's veterans to work in the shadows of the Cold War. They entered US service in secret, working in Occupied Japan and the United States to produce materials for clandestine operations. Most former members of Noborito worked after the war to rebuild Japan. Scientists and technicians who had worked on military projects in the war years contributed to Japan's postwar economic recovery and industrial expansion.

Outside of Japan, there has been no detailed history of the Japanese Army's institute for spy gear and special weapons. Excellent Japanese intelligence histories have appeared in recent years. Little exists in English, apart from some books and articles covering some aspects of Noborito's bombing balloons.

My purpose in writing this book on Noborito is to present to readers around the world a history of a Japanese intelligence organisation comparable to those of the United States, Great Britain, and other powers of that time. I have searched American archives and read the accounts of Japanese veterans for information, but much about Noborito remains in the shadows. The Japanese Army ordered the destruction of the institute's documents at the end of the Second World War. The US government has released, to my knowledge, no documents concerning its postwar projects with Noborito veterans in the Cold War. This book is the first in English on Noborito. May it serve as a foundation for future, even fuller accounts.

ACKNOWLEDGEMENTS

Many people have helped me over the years that I spent gathering materials and writing this book. I began to gather materials and write soon after publication of *The Shadow Warriors of Nakano*, my 2002 book on the IJA Nakano School for intelligence officers and commandos. My intent was to follow the Nakano book in short order with a history of the related IJA Noborito Research Institute. I reviewed in 2002 a Noborito veteran's memoir, then sketched a brief history of the institute in an article published in 2004.[1] As it happened, however, I only found the time to write this book after retiring from my career. I regret that some of the people who kindly helped me passed away before I could acknowledge them.

In Japan, staff of the National Diet Library, the National Institute for Defence Studies, and the Japan-Korea Cultural Foundation were helpful in my searches for articles, books, and documents. I am grateful to the Defunct Imperial Japanese Army Noborito Laboratory Museum for Education in Peace for permission to use a number of the museum's photographs and, in particular, to staff members Shiina Maho and Tsukamoto Yuriko for their help and guidance. I am grateful to my late friend and colleague, Yukie 'Mary' Sunaga, who many years ago gave me numerous Japanese press clippings on the Japanese Army.

In the United States, staff at the National Archives and Record Administration in College Park, Maryland, helped me find relevant government documents. Staffers of the Library of Congress, the MacArthur Memorial Archives and Library, and the US Army Center of Military History also helped me find useful materials. The Gemological Institute of America kindly sent me articles on Noborito's chief counterfeiter, who made a postwar career of gem authentication.

I offer my thanks to Tara Moran and Harriet Fielding at Pen and Sword, who commissioned this work and managed the manuscript's trans into a book, and to the company's Jon Wilkinson for designing an impressive cover. I also appreciate the work of those at SJmagic DESIGN SERVICES who handled the proof corrections and typesetting.

I am grateful to the many people who offered me advice and encouragement over many years, including the late Duval Edwards and Grant Ichikawa. For those who sent me materials and read draft chapters, I thank Marc Abramson, Akiba Ryota, Ross Coen, Charles Daly, Ron Drabkin, Robert Eldridge, Kotani Ken, Amanda McVety, and Richard Samuels. I owe a great deal to my wife, Rie, for her help and support.

Any errors in the book are mine alone.

SOME NOTES ON NAMES

Japanese, Korean, and Chinese names appear in this book in their traditional order, the family name coming before the given name. Chinese names appear as they did in the Second World War. Alternate versions for some names are found between parentheses in the index.

Place names are those of the years in which this story takes place. Korea and Taiwan in their years under Japanese rule were known respectively as Chosen and Formosa. Seoul was known as Keijo, Taipei as Taihoku, and so on. I use these and other past names of those years to avoid anachronisms. To refer to Nanking in the Second World War as Nanjing or to Fusan under Japanese rule as postwar Pusan, let alone today's Busan, to offer two examples, would be like referring to the Byzantine capital of Constantinople as Istanbul prior to the city's fall to the Turks in 1453.

As for the names of Japanese organisations, many different English versions of them have appeared over the years. One example is the 9th Army Technical Research Institute (第九陸軍技術研究所, *Dai Kyu Rikugun Gijutsu Kenkyujo*), commonly known as the Noborito Research Institute (登戸研究所, *Noborito Kenkyujo*). Official documents produced during the war and Japan's subsequent occupation referred to Noborito as an institute, laboratory, research centre or research institute. The same holds true for other organisations. A good example is the *tokumu kikan* (特務機関), a Japanese military intelligence unit. The organisation has commonly appeared in English translation as special service agency (SSA), special service organ or, simply, *tokumu kikan*. Looser translations have included "Secret Services Agency" and "Japanese Military Mission".

Things are no easier for the parts of Japanese organisations. The administrative units *kyoku* (局), *bu* (部), *ka* (課), and *han* (班) are often translated in English as bureau, department, section, and group.

However, there has been a myriad of translations for such seemingly simple terms.

For standard translations of Japanese organisations, as well as military and technical terms, I have relied on authoritative reference works of the Second World War, including the US War Department's *Handbook on Japanese Military Forces* (1944) and the US Navy's *Japanese Military and Technical Terms* (1945). Reports of the Allied Translator and Interpreter Section (ATIS) have also been helpful.

As a final note on names, mistakes are found in many official documents from the war and the period of Japan's occupation. Some are errors of typing. Other are misreadings of the Chinese characters used in Japanese. I have corrected such mistakes as necessary without flagging them in an effort to keep endnotes to a minimum.

ABBREVIATIONS AND ACRONYMS

AFFE	Army Forces, Far East
AFPAC	Army Forces, Pacific
AGS	Army General Staff
AGT	Association of Japan Gem Trust
AOAH	Army Ordnance Administrative Headquarters
ASRI	Army Scientific Research Institute
ATH	Army Technical Headquarters
ATIS	Allied Translator and Interpreter Section
ATRI	Army Technical Research Institute
BCOF	British Commonwealth Occupation Force
BW	biological warfare
CCD	Civil Censorship Detachment
CCF	Chinese Communist Forces or Communist Chinese Forces
CCRAFE	Combined Command for Reconnaissance Activities Far East
CCRAK	Combined Command for Reconnaissance Activities, Korea
CI	counterintelligence
CIC	Counter Intelligence Corps
CMH	United States Army Center of Military History
COMINT	communications intelligence
CPV	Chinese People's Volunteers
CW	continuous wave
CWS	Chemical Warfare Service
DA	Defence Agency
DPRK	Democratic People's Republic of Korea
DRV	Democratic Republic of Vietnam
FAO	Food and Agriculture Organisation
FBIS	Foreign Broadcast Information Service, Foreign Broadcast Intelligence Service
FHO	Fremde Heere Ost (Foreign Armies East)

GHQ/SCAP	General Headquarters, Supreme Commander for the Allied Powers
GIA	Gemological Institute of America
GPSO	Government Printing Supplies Office
GSDF	Ground Self-Defence Force
IJA	Imperial Japanese Army
IJN	Imperial Japanese Navy
IMA	Imperial Military Academy
JACAR	Japan Centre for Asian Historical Records
JSPS	Japan Society for the Promotion of Science
KGB	*Komitet Gosudarstvennoy Bezopasnosti* (Committee for State Security)
KLO	Korea Liaison Office or Korean Liaison Office
KPA	Korean People's Army
LCP	Laboratory of Comparative Pathology
MAF	Ministry of Agriculture and Forestry
MIS	Military Intelligence Service
MITI	Ministry of International Trade and Industry
MOC	Ministry of Communications
MPD	Tokyo Metropolitan Police Department
MSDF	Maritime Self-Defence Force
NACP	National Archives at College Park, MD
NAVTECHJAP	Naval Technical Mission to Japan
NCO	non-commissioned officer
NIBS	Nippon Institute for Biological Science
NKA	North Korean Army
NKPPC	North Korea Provisional People's Committee
NSA	National Security Agency
NYK	Nippon Yusen Kabushiki Kaisha
O C Sig O	Office of the Chief Signal Officer
OC	Office of Censorship
OGGC	Office of the Governor-General of Chosen
OSRD	Office of Scientific Research and Development
OSS	Office of Strategic Services
Q	Quartermaster
QMTID	Quartermaster Technical Intelligence Detachment
PRC	People's Republic of China

ROC	Republic of China
ROK	Republic of Korea
SAD	Special Activities Division
SOE	Special Operations Executive
SSA	special service agency
SSGHQ	Security Service General Headquarters
Tec 4	Technician 4th grade
TIC	Technical Intelligence Company
TID	Technical Intelligence Detachment
TLID	Technical Liaison and Investigation Department
USAMGIK	United States Army Military Government in Korea
USSBS	United States Strategic Bombing Survey
WDC	Western Defense Command
WDIT	War Department Intelligence Targets Section
WVF	World Veterans Federation

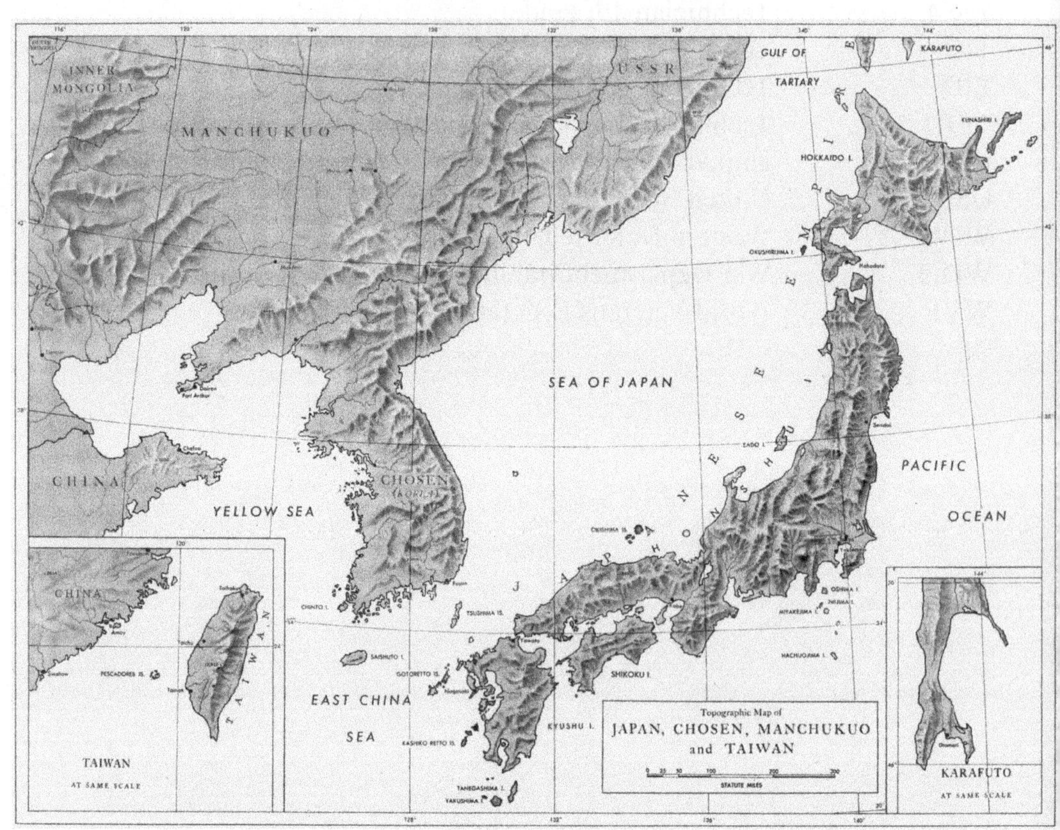

US military topographic map of Imperial Japan, showing the four Japanese main islands (Hokkaido, Honshu, Shikoku, Kyushu), Chosen (Korea), Formosa (Taiwan), Karafuto (southern Sakhalin), and the puppet state of Manchukuo (Manchuria). (Courtesy of the US National Archives)

CHAPTER 1

BUILDING NOBORITO

> I made Shinoda, a colonel at the time, the director of the [Noborito Research] Institute, which made all those materials. Among them were various things, from passports and counterfeit bank notes to, for example, bombs shaped like lumps of coal that, placed among lumps of coal, would explode.
>
> – Major General Iwakuro Hideo[1]

Armies had long fought on fields of battle and in the shadows behind enemy lines. Advances in science and technology since the start of the Industrial Revolution led to deadlier weapons and more sophisticated equipment for clandestine operations. Repeating rifles replaced muskets. Chemicals replaced lemon juice and vinegar in secret inks. In modern armies, new organisations conducted research and development to produce weapons, equipment, and materials for fighting and spying.

Such advances brought devastating weapons to the battlefields of Europe in the First World War. In 1903, the Wright brothers had flown the world's first motorised aeroplane a few feet above a coastal sand dune in North Carolina. Before the Armistice of November 1918 ended the fighting, pilots were strafing battlefields, dropping bombs, and downing enemy aircraft. Gottlieb Daimler and others had developed motor vehicles around the turn of the twentieth century. In the Great War, armoured tanks brought a new form of terror to the battlefield. Chemical scientists and engineers developed both modern fertilisers and poison gas that killed at Ypres and other battlefields over 90,000 men and wounded well over a million.[2] The horrible new weapons resembled works of science fiction but were all too real. The marauding submarine *Nautilus* in Jules Verne's *Twenty Thousand Leagues Under the Sea* (1870) and the gas attacks in H.G. Wells's *War of the Worlds* (1898) were now real.

Nations also developed new ways of secret warfare. Imperial Germany, a leader in the life sciences, conducted clandestine operations to kill or sicken animals to deny them to the Allied powers. Beginning in 1915, German agents attacked animals in the United States with anthrax and glanders. They spread anthrax among cattle and sheep to disrupt food supplies. Glanders was for horses and mules used for hauling military equipment and supplies.[3]

Progress in science and technology also brought advances in secret inks and other tools of espionage. From Germany's chemical industry came new inks by which German agents hid messages. Earlier inks had appeared with the application of heat. The Germans advanced the art of spying with chemical ones that remained invisible even when heated. In addition, agents carried their ink not in bottles, which could attract suspicion, but impregnated in clothing. Once safe, agents would soak the garments in a certain liquid to bring forth the writing.

Government organisations established organisations for intelligence and counterintelligence. The US War Department in that era developed an eighth section in military intelligence (MI-8) and other units of its Military Intelligence Division (MID). Under Herbert O. Yardley, a former code clerk in the State Department, MI-8 compiled codes and broke those of foreign powers. With British help, Yardley set up a laboratory to develop secret inks and detect those of the enemy. Military intelligence employed diverse talents, from chemists adept at using the iodine vapor bath for revealing secret inks to criminals skilled in forgery.[4]

The Western powers continued after the war to conduct clandestine operations and to develop spy gear and weapons. International conflicts in the 1930s led to innovations and new organisations. After the German invasion of Poland in September 1939 ignited the Second World War in Europe, the British government established the SOE in secret in July 1940. Station 12, an SOE outstation, produced unconventional weapons for British special operations. After Japan attacked the United States in December 1941, striking American warships at Pearl Harbor in Hawaii and invading the US territories of Guam, the Philippines and Wake Island, Washington established in June 1942 the OSS. Its Research and Development Branch (R&D) developed spy gear and special weapons.

Former OSS officer E. Howard Hunt recalled years later R&D's 'amazing array of weapons and gadgets,' including explosive powder 'packaged to look like Chinese flour,' a 16mm camera concealed in a matchbox, portable kayaks, and counterfeit documents and currency 'printed by R&D to the most exacting and up-to-date standards.' OSS veteran Roger Hall recalled that while in training he saw explosive devices among R&D's 'dozens of gadgets'. Impressed, he wrote: 'Most of them were hellishly clever, and all were effective.'[5] Organizations such as SOE's Station 12 and OSS/R&D gave Allied intelligence officers and commandos the means to destroy, spy, travel, and survive behind enemy lines in the Second World War.

Japan's Army Moves to Close the Gap

Imperial Japan, a member of the Allied coalition in the First World War, fought in Asia against the German Empire largely without fielding the latest battlefield weapons or advanced equipment for clandestine operations. In contrast to Europe, with its deadly innovations, the fighting in Asia was small in scale and conventional in character.

The Western powers led Japan in fielding new weapons and developing new organisations. In 1916, Britain began construction of a facility at Porton Down for research and development in chemical warfare. The US Army established its Chemical Warfare Service (CWS) several months before the 1918 Armistice. The Japanese Army at that time had no equivalent organisation. Nor did Tokyo match Washington or its other allies in establishing an organisation to apply technology to intelligence operations. The Japanese Army fought the war without a unit on a par with the US Army's MI-8.

Tokyo, Asia's only great power in a world dominated by the West, was determined to close the gap. In 1919, the Japanese Army established the Army Technical Headquarters (ATH) and its subordinate ASRI. The ATH gathered domestic technical information through its research section and foreign information from military technical officers, special service agencies (SSAs), Japanese military units, and 'special means.' The ASRI conducted research on weapons for both overt and clandestine warfare. In June 1941, the ATH absorbed the ASRI; the larger ATH organisation included nine numbered institutes. In 1942, the War Ministry combined its Weapons Bureau, the Army Ordnance Headquarters, and the ATH into the Army Ordnance Administrative Headquarters (AOAH). Under the new headquarters, each numbered institute received a new designation: army technical research institute (ATRI). The AOAH added a tenth, and final, ATRI in 1944.

Of the ten institutes under the AOAH, the 9th ATRI stood apart. Its mission was to research and develop spy gear and special weapons. The other institutes conducted research and development for battlefield equipment and weapons. Located southwest of Tokyo at Noborito, in what is today the Tama Ward of the city of Kawasaki in Kanagawa Prefecture, the 9th ATRI was known as the Noborito Research Institute.[6]

The formidable challenge of spying and operating against the Soviet Union, Imperial Japan's foremost adversary, drove the Japanese Army to modernise its military intelligence activities. In 1934, the Soviet Union sealed its border with the Japanese puppet state of Manchukuo, ending border crossings by inhabitants on both sides. The Soviets also increased the number of border garrisons, added sentry posts, and conducted frequent canine patrols. Japanese intelligence activities along the border thus 'came to a standstill at the end of 1934,' according to a postwar assessment. In 1937, Moscow resettled Koreans and other residents from areas in the Soviet Far East along the borders of Chosen and Manchukuo to the Soviet interior to deprive the Japanese of potential agents for intelligence operations.

Japan also saw China as a challenge. Tokyo had attempted to subdue Nationalist China by force of arms after a clash in July 1937 between Japanese and Chinese military forces at the Marco Polo Bridge in Peking. The local incident took place in the midst of Tokyo's manoeuvres to detach and control northern China after having gained control of Manchuria and turned it into Manchukuo. In the months after the clash at the Marco Polo Bridge, Japan took control of much of the Chinese coast but lacked the military resources – most of which were for an anticipated future conflict with the Soviet Union – to conquer all of China. The Japanese Army needed to find other means to end China's resistance.

The Japanese Army responded to these developments by acting, in the language of a postwar report, 'to break the deadlock in espionage activities by modernising and systematising intelligence activities'.[7] Creating new units to bolster intelligence operations was part of it. The Japanese Army created in November 1937 within AGS Second Bureau a dedicated unit for secret operations: 8th Section. The IJA Kwantung Army in Manchukuo upgraded its SSA network, establishing a Kwantung Army Headquarters Intelligence Section, turning the Harbin SSA into its intelligence headquarters and subordinating the SSA in Dairen and other agencies in Manchukuo to it. To build its counterintelligence (CI) capabilities, the Army created in 1937 a new unit: the Yama Agency. A secret even within the IJA, the Yama Agency intercepted letters, tapped telephone lines, including those of foreign embassies, and engaged in other CI activities. In 1938, the Army established the Nakano School to train intelligence operatives.[8]

The Japanese Army took measures to train and equip field units. In November 1938, the Army sent Colonel Shinoda Ryo, an engineer officer in charge of the ASRI laboratory for clandestine warfare materials, and the technician Ban Shigeo, his principal assistant, on a month's tour to train IJA units in Manchukuo. Shinoda and Ban arrived in Hsinking, the capital, with two briefcases and three large trunks of secret documents and materials. Escorting them were officers of the Kempeitai, the fearsome military police responsible for both regular policing and for CI activities. After reporting the next morning to senior intelligence officers of the Kwantung Army's Kempeitai Headquarters, Shinoda briefed clandestine operations – including intelligence tools and methods of detecting clandestine communications – to about 40 Kempeitai officers. Shinoda and Ban then toured Kwantung Army SSAs in Harbin, Mukden, and other areas before returning to Tokyo.

In September 1939, Shinoda and Ban returned to Manchukuo for another six weeks to train officers and to take technical requirements for local operations. In February 1940, Shinoda sent Ban and another technician to the Nanking headquarters of the China Expeditionary Army and to IJA units in Shanghai to demonstrate Noborito equipment and weapons, including secret inks and explosives disguised as bricks. In Shanghai, the two technicians procured via the Kempeitai bottles of whiskey, tins of hard sweets, and other Chinese items as samples for projects in Noborito.[9]

Leaving Tokyo for Noborito

The ASRI at its start in 1919 comprised two sections: one for physics and the other for chemistry. Gunpowder and explosives were the focus of the second section. A restructuring in 1925 replaced the two sections with three departments. Department 1 continued the work of the first section in physics. Department 2 followed the course of the second section in explosives. The new Department 3 was for basic research in chemical warfare. In 1932, with research on explosives transferred elsewhere, the remaining two departments focused on emerging technologies. Members of Department 1 worked in such areas as electric-wave weapons, including a death ray. Those in Department 2 worked on offensive and defensive aspects of chemical warfare and on poisons. The ASRI's emphasis at that time was on chemical warfare. Colonel Satake Kinji, an engineer officer involved in weapons R&D, wrote after the war that, although the ASRI on paper had seemingly equal divisions between the physical and chemical sections, offensive and defensive work on poison gas accounted for most of the institute's activities.[10]

In April 1927, the ASRI started in Department 2 a laboratory under Shinoda Ryo, then an army captain, to develop equipment for clandestine operations. What he and his assistant Ban would later call the 'Army's cradle for clandestine warfare materials,' the laboratory in its early days comprised Shinoda, Ban, and two other assistants. In working to meet the Soviet and Chinese intelligence challenges, the staff over time outgrew the ASRI's headquarters of red brick in what is today the Shinjuku area of Tokyo. In 1939, by then a colonel, Shinoda moved his laboratory to what became that year the ASRI's Noborito Branch Depot, originally established in 1937 as an ASRI laboratory for electric-wave research.[11]

The new site was located 'on quiet heights overlooking the Tama River', Shinoda and Ban recalled years later. The Odawara Line, with its Tokyo terminal station near the ASRI, connected the institute's headquarters and its Noborito branch. The grounds comprised approximately 36 hectares and came to include approximately 100 buildings and a number of places for testing explosives and other materials. Shinoda assumed command of the growing facility soon after moving his laboratory there. Major General Iwakuro Hideo, a driving force in those years to modernise Japanese military intelligence, claimed credit. He told an interviewer years after the war that he had established the Nakano School to train military intelligence personnel, launched Noborito to make the military's spy gear and special weapons, and picked Shinoda for the institute's director.[12]

Deadly Work in a Pleasant Atmosphere

Noborito was enveloped in secrecy and engaged in deadly work. Oshima Yasuhiro, a Noborito technician, recalled years after the war that the Kempeitai had conducted a strict security check of his background.

When he reported for duty, Oshima was told to keep Noborito's secrets and was warned that trying to quit the institute would be seen as an act desertion, punishable by execution.[13]

Officers, technicians, and other employees at the ASRI and Noborito enjoyed working conditions and a general atmosphere better than what most in the Japanese Army experienced. One technician, who had joined the ASRI following his discharge from obligatory military service in 1938, found that Noborito salaries were higher and, 'because we were all researchers, the atmosphere was liberal there'. Seki Koto, one of Second Section's two female typists, who had started working there from the age of 15, enjoyed lunches and games of ping-pong with the other young woman in her section.[14]

Shinoda was the right type of officer to create such an atmosphere. Ban recalled his superior's character as 'more that of a scientist than a soldier.' An expert in cellulose chemistry who had studied in Great Britain, Shinoda impressed his subordinate as 'the model of an English gentleman and a calm technocrat'. Kawamoto Kazuko, who started in 1941 at the age of 16 handling general office duties before tracking shipments in the weapons section of the accounting department, recalled Shinoda's 'kind words' to her when she would serve him tea. Yuhara Hiroo, a technical captain who worked in First Section on communication devices, described the atmosphere there as 'an extension of a university laboratory,' where superiors and their subordinates 'were joined together by a common understanding as technicians.'[15]

Noborito's researchers read widely as they developed spy gear and special weapons. The ASRI practice of researching domestic and foreign technical literature continued atop the quiet heights. Researchers would start each research project by spending the first three to six months poring over domestic and foreign publications. A startling book from America – Yardley's *Black Chamber* – offered intelligence pointers. The author, who had worked after the war in the State Department's Cipher Bureau, revealed in his 1931 book that the bureau had broken codes, primarily Japanese ones, until its closing in 1929. *The Black Chamber*, which hit bookstores in Japanese translation the same year as the original, also revealed the use of iodine vapor in detecting secret inks and other details on clandestine communications.

Ban took a 'keen interest' in Yardley's book, as well as in the British analytical chemist Alfred Lucas's *Forensic Chemistry and Scientific Criminal Investigation* and a book on analytical chemistry by the German chemist Richard Berg. Noborito researchers also turned to fiction for inspiration. Shinoda and Ban wrote years later: 'Around the time when we were getting involved in research, we began to pioneer an unknown world with nothing to consult other than spy novels and movies.'[16]

As Japan lagged the Western powers by the end of the Great War in conventional military science and technology, so Noborito trailed foreign organisations in spy gear and special weapons. Shinoda and Ban assessed after the war that their organisation had compared poorly to counterparts in the Soviet Union, Nazi Germany, Great Britain, and the United States. They described Japan as suffering from a 'marked gap' over the years in budget, effort, experience, and scope of activities.

Noborito narrowed, perhaps even closed, the gap. Working with the Kempeitai from the time of the Manchurian Incident in 1931, Noborito improved the equipment of Japan's military police to world standards, enabling the Japanese Army at home and abroad to detect and eliminate foreign plots. Shinoda and Ban wrote with pride that Noborito helped to make 'the name of the Japanese Kempeitai feared in the world.' Just before the war ended, they claimed, Noborito managed 'at last to systematise clandestine warfare materials across the board.'[17]

Noborito officers and technicians enjoyed support from on high. The institute had by far the largest budget of the 10 ATRIs. In the last year of the Second World War, Noborito had a budget of approximately 6.5 million yen. With total funding for all 10 institutes that year totaling roughly 35 million yen, Noborito alone accounted for close to a fifth of all ATRI funding. With such backing, Shinoda's original ATRI staff of three laboratory assistants grew at Noborito to a workforce of nearly a thousand people.[18]

Products

Noborito equipment and materials, according to Ban, covered five main categories:

1) intelligence
2) counterintelligence
3) clandestine operations
4) propaganda
5) Kempeitai

Secret inks were among the items that Noborito produced to communicate intelligence. Radio direction-finding equipment to pinpoint enemy agent transmissions and devices for recording conversations were among the CI products. Noborito also made materials to 'dust' for fingerprints and to open and seal intercepted mail without leaving a trace. For clandestine operations, Noborito developed disguises and material to counter canine patrols by distracting or paralysing dogs. The institute concocted poisons, some added to coffee, whiskey, and other food and drink. Explosives came in canned food and other guises. Sugita Masami of Noborito's Second Section recalled packing explosives disguised in cans with the label printed in Cyrillic indicating crab meat, the devices evidently meant for operations against the Soviets. Under the code name Se-go, Noborito developed a propaganda vehicle equipped with printing equipment and materials, loudspeakers, and other devices. Much of what Noborito produced went to the Kempeitai. The Kempeitai officers operating under the Kwantung Army, reportedly 'head and shoulders' above other units of the IJA military police in intelligence capabilities, were in particular need of Noborito's technology for operations against the Soviet Union.[19]

Supporting the Kempeitai and Yama Agency

In counterintelligence, Noborito technicians not only developed CI tools but, at times, fielded requests to work on particular problems. In one case, members of Ban's Group 1 in Noborito's Second Section applied a solution to peel away a postage stamp from an envelope without damaging the stamp, revealing secret writing on the stamp's reverse side. The discovery led the Kempeitai to arrest those at the sending and receiving ends of the intercepted letter.

The Japanese Army highly appreciated the work of Shinoda's men even before his laboratory made the move to Noborito. Soon after the Japanese Army fought its way into Shanghai in 1932 in the First Shanghai Incident, Kempeitai officers seized a couple metres of striped cloth used for making dress shirts. Suspicious but unable to evaluate the material in the field, they sent the material to Tokyo Kempeitai Headquarters, which in turn passed the cloth to the ASRI. Shinoda had Ban tested the cloth for secret writing. Working on his first 'urgent' assignment, Ban tested a portion of the cloth with an iodine solution, which worked on a starch solution in the cloth to reveal orders of the Communist Party of China. Ban's success earned praise the next day when Shinoda personally briefed Field Marshal Prince Kanin, the chief of the Army General Staff, as well as other top AGS officers.

In another case, Ban fielded in December 1941 a request from the Kempeitai in Formosa to examine a document suspected of containing a message written in invisible ink. Ban tasked two experienced technical officers – Hase Michio and Arikawa Shunichi – and a technician to work with him on the problem. The substance, which Ban described as a 'high-class secret ink' from the United States, was a tough nut to crack. After consulting superior officers Shinoda Ryo and Yamada Sakura, Ban called in Dr Ueno Shigezo, a specialist in dye chemistry at Tokyo Technical College. After several weeks of experiments, the group succeeded in uncovering the message. Noborito reported the group's success to the AGS and Kempeitai Headquarters.[20]

Shinoda's people also supported the new counterintelligence organisation hidden in the War Ministry's Military Affairs Bureau: the Yama Agency. Few even within the Army knew of Yama's existence. In 1937, Major General Anami Korechika, director of the Military Affairs Bureau, had ordered Lieutenant Colonel Akikusa Shun, a Soviet expert, and Major Fukumoto Kameji, a Kempeitai officer, to establish a 'scientific CI organ.' The two worked with Major Utsunomiya Naokata, another intelligence officer, to stand up that same year a secret organisation. Akikusa became the new unit's first chief; Fukumoto served as his deputy. Officers of the Kempeitai and the Nakano School came to staff the new unit. Shinoda's men developed devices for Yama, including recording equipment. Maruyama Masao's Group 5 in Second Section developed a spy camera used, in one instance, in the surveillance of Yoshida Shigeru. The former ambassador to Great Britain, arrested in April 1945 after a period of surveillance as part of a group of influential Japanese seeking an early end to the war, later became one of Japan's most notable postwar prime ministers.[21]

Noborito also contributed to discovering the illegal radio transmitters of enemy intelligence agents. In one case, the Kwantung Army's Kempeitai used equipment to detect radio signals to uncover in Hsinking a Soviet intelligence operation operating from a radio shop. The Kempeitai succeeded after forming a special unit for finding illegal radio transmissions. Their feat, according to Shinoda and Ban, was due to the Kempeitai's 'people, equipment, and organisation.' It was partly due to equipment from Noborito. Shinoda and Ban claimed years later that military investigators had arrested 'more than 215 spies, with their radios, in Japan, Chosen, Manchuria, China, French Indochina, Thailand, Malaya, and Burma' in the period between the Manchurian Incident in 1931 and the end of the Second World War in 1945.[22]

Breaking a Soviet Spy Ring

Noborito also played a role in breaking a Soviet spy ring that operated in Tokyo for years. Richard Sorge, a Russian-born German veteran of the Great War, had entered Japan in 1933 to gather intelligence for the Soviet Red Army. Working as a correspondent for a German newspaper and posing as a Nazi, he had succeeded in cultivating the German ambassador and other members of the German Embassy in Tokyo, recruiting a member of Prime Minister Prince Konoe Fumimaro's brain trust, and developing other sources of information. In 1941, Sorge warned his superiors that Germany was planning to invade the Soviet Union. He also reported the ultimate decision of a wavering Japan to forego aiding its German ally by invading the Soviet Far East in favour of seizing American, British, and Dutch colonies in Southeast Asia for needed petroleum and other raw materials.

In the spring of 1941, the Yama Agency launched a mobile unit to detect illegal radio transmissions. Noborito technical officer Takano Yasuaki had developed the equipment. His 'direction finder for illegal radio detection' comprised an all-wave receiver with a cathode-ray tube display. The equipment occupied the rear of a vehicle, with loop antennas protruding through the roof. With three such vans, members of the Yama Agency drove routes around Tokyo to triangulate the position of radio transmissions. Takano later recalled that the mobile unit's members often visited him for guidance as they tested the vehicles and readied them for operations.

Within months, the Yama Agency used Takano's equipment to catch Sorge. Tipped to illegal radio broadcasts by the Communications Ministry, which conducted fixed-site radio direction finding, the Yama Agency went into action. Narrowing down the transmission area, they put part of western Tokyo under surveillance for twenty-four hours each day. On 14 October 1941, they pinpointed the transmission source. Inside the residence was Max Clausen, Sorge's radio operator.

The Yama Agency passed their intelligence to the Kempeitai, which informed the Tokyo Special Higher Police (*Tokko Keisatsu*). The latter organisation, which reportedly had not suspected Clausen, immediately began to watch his house. The next day, detectives observed Sorge and another member of the ring enter the house. The Special Higher Police arrested the three spies on 18 October. The police received credit for the case. The role of the IJA's Yama Agency, and its Noborito equipment, remained hidden for years.[23]

Equipping the Nakano School for Clandestine Operations

Noborito also worked with the Nakano School. The IJA training centre for intelligence officers and commandos would call on Noborito to develop equipment. The institute responded by delivering pistols disguised as cigarette lighters, incendiary bricks, explosives disguised as lumps of coal, poisons, disguises, and other devices. Noborito would dispatch personnel to the school to instruct the students, who would tour Noborito as part of their education. Some of the graduates before leaving on missions would go to the institute for hands-on training in handling special equipment. Shinoda himself instructed Nakano students on foreign weapons and military fortifications. Ban and Takano also taught there. As a postwar history of the Nakano School put it, the relationship with Noborito was 'a close and inseparable one.'[24]

Major Kuwahara Takeshi, a graduate of both Imperial Military Academy (IMA) Class 52 (1939) and Nakano School Class 2B (1944), gave proof of that relationship. A military operative in Burma with the Japanese Army's Hikari Agency, Kuwahara recalled the close relations between Nakano and Noborito: 'Our proving unit would conduct various tests and try out what the Noborito Research Institute made on a trial basis.'[25]

Personnel and Organisation

Commissioned engineer officers, technical officers, and technicians directed research, development, and production at Noborito, which comprised three main sections, each with several R&D groups and areas of concentration, and a fourth section for testing and production. Below is an outline of Noborito's leadership and organisation. The ranks and titles given are those as of the end of the war. Some project areas of each group are listed between parentheses:

9th Army Technical Research Institute (Noborito Research Institute)
Director: Lieutenant General Shinoda Ryo

First Section: Major General Kusaba Sueki

- General Affairs Group: Captain Nakamoto Toshiichiro
- Group 1: Major Takeda Teruhiko (bombing balloons, propaganda vehicles)
- Group 2: Major Takano Yasuaki (special radios, radiosondes)
- Group 3: Technician Sasada Sukesaburo (death ray)
- Group 4: Major Otsuki Toshiro (artificial lightning)

Second Section: Colonel Yamada Sakura

- Group 1: Major Ban Shigeo (clandestine communications, counterintelligence equipment, weapons for clandestine operations)
- Group 2: Major Murakami Tadao (poisons)
- Group 3: Major Hijikata Hiroshi (poisons for clandestine operations)
- Group 4: First Lieutenant Kuroda Chotaro (weapons for clandestine operations against animals)
- Group 5: Major Maruyama Masao (spy cameras, microdots, copying equipment)
- Group 6: Major Ikeda Yoshio (weapons for clandestine operations against plants)
- Group 7: Major Kuba Noboru (weapons for clandestine operations against animals)

Third Section: Colonel Yamamoto Kenzo

- North Group: Major Ito Kakutaro (paper manufacturing)
- Centre Group: Major Okada Masayuki (analysis, forensics, printing ink)
- South Group: Major Kawahara Hiroma (plate making, printing)

Fourth Section: Colonel Hatao Masao

- Major Natsume Isuo (production, supply, training for items developed in First and Second Sections)[26]

Career information for the institute's director and four section chiefs:

- Lieutenant General Shinoda Ryo, director, Noborito Research Institute: engineer officer, IMA Class 26 (1914); graduated in 1922 from Tokyo Imperial University, later receiving from the university a doctoral degree in engineering; sent to Great Britain to study, 1929-1932, at the University of London.[27]
- Major General Kusaba Sueki, director, First Section: engineer officer, IMA Class 32 (1920), assigned on receiving his commission to the ASRI; studied physics at Tokyo Imperial University, 1923 – 1926; posted to Germany, 1935 to 1937, for studies; director, ASRI Noborito Laboratory, from December 1937 to March 1939.[28]
- Colonel Yamada Sakura, director, Second Section: technical officer with civilian background; bachelor (1923) and doctoral (1942) degrees in engineering, applied chemistry, Tokyo Imperial University; joined the ASRI in 1923 on graduation from university; director of Noborito's Second Section from August 1943 to war's end; author of a 1936 book on chemical weapons in which he dismissed the 1925 Geneva Protocol's ban on those weapons as 'no more than a scrap of waste paper.'[29]
- Colonel Yamamoto Kenzo, director, Third Section; intendance officer; Army Intendance School (1921); sent in 1921 to Soviet Far East in Japanese Army's Siberian Expedition; sent in 1933 to Kwantung Army, studied local topography, currencies in Manchukuo; assigned in 1938 to 7th Section (China) in AGS Second Bureau; given concurrent assignments in 1939 to 9th ATRI and AGS 8th Section.[30]
- Colonel Hatao Masao, director, Fourth Section: engineer officer, IMA Class 34 (1922); later sent to Tokyo Imperial University for further studies in electrical engineering; posted in 1936 to Germany as an ATH officer; assigned to the ASRI in 1939. Hatao assumed command of Fourth Section in February 1943 after serving for a time as director of Second Section.[31]

Prominent specialists from academia and industry also worked for Noborito, contributing as contract researchers their expertise in various fields. At least 64 civilian academics, scientists, and engineers served as consultants or contract researchers. Some also worked at other ATRIs. Among them was Dr Yagi Hidetsugu of Tohoku Imperial University, inventor with colleague Uda Shintaro of the Yagi-Uda antenna, used to transmit signals over long distances in such applications as radio and television.[32]

Growth of a Secret Organisation

The growing Noborito Research Institute was a secret within the Imperial Japanese Army, as it was on the outside. Established at first as a place for electric-wave research, the ASRI Noborito Laboratory grew to become the Noborito Branch Depot and then the Noborito Research Institute. When Shinoda took command in September 1939, Noborito had added a section for biological warfare, then a third section for counterfeiting and document forgeries. A classified IJA document from that month discretely described First Section as engaged in 'electric-wave research' and Second Section as involved in 'special scientific materials.' The internal document made no reference at all to Third Section.[33]

CHAPTER 2

DEVELOPING A DEATH RAY

> The era of fire arms and artillery using explosives has reached an impasse. We are entering an era of electricity. We should use electricity on its own as a weapon, such as electric death rays and dazzling beams.
>
> – Lieutenant General Tada Reikichi, chief, Army Technical Headquarters[1]

New Weapons for the Next War

Military leaders of the great powers, concluding from the introduction of poison gas and other new weapons on European battlefields in the First World War that rival nations would produce advanced arms in future wars, took steps to develop their own cutting-edge weapons. In some cases, development took the form of improvements to existing weapons. Military aircraft, for example, flew faster and farther with successive technical developments. There were also new areas of science and technology to explore. Electric waves were one such area. Scientists and engineers were applying such waves in the developing fields of radio, radar, and television. In addition to using them for communication and the detection of objects over long distances, military planners also thought of turning electric waves into offensive weapons. Colonel Satake Kinji explained: 'The various new weapons that appeared in the previous world war gave birth after the war to visions of various new weapons. The death ray was one of them.'[2]

Death Rays in Popular Culture and Foreign Reports

Scientific and technological advances in the nineteenth century led writers to imagine weapons that emitted powerful rays of energy.

The British novelist H.G. Wells wrote in his novel, *The War of the Worlds*, published near the end of that century, of invaders from Mars wielding heat-rays from which came an 'almost noiseless and blinding flash of light' to incinerate their targets. Two German writers, Max Seydewitz and Kurt Karl Doberer, published in London in 1936 a speculative work on death rays as future weapons.[3] Ray guns of one kind or another became common in tales of science fiction printed in pulp magazines in the first half of the twentieth century. Buck Rogers, Flash Gordon, and other heroes of the future blasted foes with beams.

The British inventor Harry Grindell Matthews claimed several years after the Great War to have invented a 'death ray' capable of stopping a motorcycle engine. The press in Great Britain and elsewhere reported his astounding claim. An article in the August 1924 issue of *Popular Mechanics* that year breathlessly wrote of the reputed invention as possessing the potential for 'entirely new fields and methods of warfare.' An accompanying captioned drawing showed 'salvos' of 'death rays' unleashed from 'batteries of transmitters' atop buildings in New York City downing attacking enemy aircraft.[4] The US weekly magazine *Time* that same year reported that such an invention would 'bring aeroplanes in flight to a full stop and send them crashing to earth.' The magazine quoted the British inventor: 'I believe that in the near future machine guns will be found only in museums.'[5]

Similar media reports appeared time and again in the interwar years. Universal Studios in the United States produced a newsreel in 1930 reporting that engineers had demonstrated '4,000,000 volt "lightning" fatal at 200 miles.' In two 1936 newsreels, Universal reported the development of a death ray to kill insects 'at a distance of more than twenty feet' and, separately, the exhibition of the 'latest "death-dealing" device', a death ray, for public viewing.[6] Even Nikola Tesla, famed for his work in developing technology for alternating current, announced in 1934 that he had invented a 'super death ray', according to a report of United Press, to kill an army of '1,000,000 men' and 'bring down a fleet of 10,000 aeroplanes at a distance of 250 miles.' Tesla cautioned that, due to the 'huge generating stations' required, it would only be practical to mount such a weapon on battleships.[7] Tesla never backed his claim with a convincing demonstration. Nor did anyone else.

The death ray appeared in those years in Japan as well as a weapon of the future. Yoshimura Misao directed for the Daito Film Company in

1936 a three-part movie series featuring a death ray. Unno Juza, Japan's pioneer in science fiction, penned in 1938 the story 'Death Ray', part of a collection of military tales published under the title *Tokyo Kubaku* [*Aerial Bombardment of Tokyo*]. In Unno's story, a Dr Todoroki, director of the Special Scientific Institute, develops a ray capable of turning a man into a charred corpse 'in an instant.' In 1940, the Tokyo publisher Nikko Shoin released in a German-Japanese bilingual edition a German work on death rays. Also that year, a publisher produced in Japanese translation the earlier book by Seydewitz and Doberer. A Japanese author included that year a section on death rays in a work of nonfiction on 'modern scientific warfare.'[8]

Turning Waves into Weapons

Such publications caught the attention of Japanese authorities. A newspaper article referring to German military research in the First World War on a death ray prompted IJA investigations. In December 1926, Dr Yagi Hidetsugu of Tohoku Imperial University, a prominent electrical engineer and inventor, spoke on invitation at the ASRI on the possibility of developing a death ray. Yagi explained that there did not exist at that time the technology required to generate sufficiently powerful electric waves for such a weapon. Furthermore, he noted that defence against such a weapon was already possible. He did say, however, that short waves seemed a promising avenue of future research.[9]

Around 1930, the Army General Staff tasked the ASRI through the War Ministry to assess the feasibility of developing an electric-wave weapon. The idea was to develop something that could use 'powerful ultrashort waves' to stop an engine by means of the resonance effect and to harm or kill people. The ASRI determined after a review of technical literature that the 'enormous energy requirements' of such a weapons system precluded its development. In 1936, the AGS again ordered a feasibility study. Officers remained interested in the potential of shutting down engines and knew by that time that ultrashort waves could cause bodily harm. Scientists in Japan and other countries had shown in experiments on animals that ultrashort waves could kill. Dr Sasada Sukesaburo, a professor at Hokkaido Imperial University, was a leading Japanese researcher in this field.[10]

Wave Work at the ASRI

In the ASRI, Department 1 was responsible for conducting research on physical weapons, including the death ray. Yamada Genzo, a technician in the department, started working on the death ray and other electric-wave projects after graduating in 1935 from Hamamatsu Higher Technical School in Shizuoka Prefecture. Such work occupied most of Yamada's wartime career, at the end of which he had reached the rank of technical major.

Driving military research on electric waves at the ASRI, according to Yamada, were two officers: Major General Tada Reikichi and Captain Satake Kinji. One ASRI officer recalled that basic work on electric waves began in Department 1 around the time that Tada took charge. It was Tada, as department director, who tasked his subordinate Satake to prepare a new facility for conducting tests with electric waves.

Tada, an artillery officer, was a member of IMA Class 15 (1903). In 1913, he graduated from the Department of Physics in Tokyo Imperial University's Faculty of Science. In 1919, the Army assigned him to the ASRI before sending him the following year to Germany for three years of studies. In 1922, Tada received orders returning him from Europe to the ASRI. In 1926, he earned a doctoral degree in engineering. He became director of ASRI Department 1 in April 1932, director of the War Ministry's Weapons Bureau in August 1934, director of the ASRI in August 1936, and director general of the ATH in March 1939. Following retirement at the end of 1940, Tada returned to service in May 1945 as president of the Cabinet's Board of Technology. A prolific writer, his writings included two wartime books: *Kokubo gijutsu* [National Defence Technology] and *Shoraisen to kagaku shinheiki* [Future Warfare and New Weapons of Science]. Tada, known for having developed artillery spotting tools, had the reputation within the Army as a 'god of land warfare weapons.'[11]

Satake, an engineer officer from IMA Class 36 (1924), had excelled at the IJA Artillery School and later studied electrical engineering at Kyoto Imperial University. Following his time at the ASRI, the IJA assigned Satake in 1941 to Berlin to serve as a technical assistant to the Army Attache Office. In 1943, he returned by U-boat to Japan and became director of Third Section at the Tama Army Technical Research Institute. The Army promoted him to colonel the next year.[12]

Tada was convinced that electricity would replace conventional explosives. He directed multiple electrical engineering projects.

Each secret project had a cover name. Those for electric-wave weapons had the designation 'Ku-go,' short for the Japanese term *kuwairiki kosen* (怪力光線, also pronounced *kairiki kosen*), literally 'rays of fantastic strength.' A military document from September 1936 shows three initial areas of research for Ku-go 'special research' projects: electric waves, radioactive rays, and shock waves. Dr Yagi consulted on electric waves, for which Satake in ASRI Department 1 was responsible. Satake was also the ASRI point man for the U-go project for investigating how to turn lightning into a weapon. For the Ku-go project, the ASRI soon abandoned radiation and shock waves for a focus on electric waves alone. Dr Yagi continued to consult, bringing foreign technical literature and giving technical guidance in monthly visits to the military researchers.[13]

Tada turned that same year for academic expertise and corporate resources to the Japan Society for the Promotion of Science (JSPS), a scientific group created in1932 under the Ministry of Education in part to join the resources of academia with those of industry and the military together for the nation's defence. Tada himself was a member of the JSPS 10th Permanent Committee, which investigated applications in electrical engineering. Satake, whom Tada had made responsible for research on electric weapons as director of Group 2 in ASRI Department 1 following his return from Germany, drafted a letter of intent that set forth the Army's goals in research on 'the generation of powerful electric waves.' The ASRI then worked with the JSPS First Subcommittee, giving both direction and research funds.

Corporate participation was essential in the testing and manufacturing of equipment. Satake found particularly valuable the 'active cooperation' of two technicians, Kobayashi Masatsugu of the Nippon Electric Corporation (NEC) and Ueno Shinichi of the Nippon Wireless Company. Such corporations were important for their human talent and laboratory resources.

Tada could obtain prototypes of vacuum tubes and other components via the JSPS, but the ASRI in downtown Tokyo lacked the requisite facilities for research, development, and testing. As a solution, the Army procured a site in Kanagawa Prefecture, approximately 18 kilometres southwest of the ASRI. Satake and Dr Kusunose Yujiro, a technician from the Ministry of Communications (MOC), found an area at Noborito, located near the Inada-Noborito Station on the Odawara Line, which terminated near ASRI headquarters. The price was high, but the War Ministry in the end secured the funds for its purchase.

Testing the Waves at Noborito

In December 1937, researchers began working at the ASRI's Noborito Laboratory. The initial staff of 18 grew to 60 by the end of April of the following year under the direction of Major Kusaba Sueki. From start to finish, much of Noborito's brief history involved the Japanese Army's quest to turn electric waves into weapons and military equipment. The laboratory formed the nucleus of what would become Noborito Research Institute's First Section. The selection of the hilltop site was deliberate, as the hill's elevation was good for emitting electric waves and measuring reflected waves. In addition to research on the Ku-go project, Kusaba directed personnel tasked with producing large vacuum tubes and conducting research on radar (designated Chi-go) and the U-go project for artificial lightning.

A key to success was in assembling talented military and civilian researchers. The ASRI sent some officers from its Tokyo headquarters to Noborito for its electric-wave projects. Yamada Genzo recalled his superior in Department 1, Captain Matsuyama Naoki, informing him of his transfer to the new facility. Yamada was a technical officer who had studied while a student at Hamamatsu Higher Technical School under Dr Takayanagi Kenjiro, the electrical engineer who had broadcast in 1926 the world's first image using an 'all-electronic' television set incorporating a cathode-ray tube. Yamada recalled that he was happy to receive his new orders for Noborito, where he would work on a project at the frontier of wireless technology. Together with Captain Matsuyama, Yamada went to Noborito to work on the Ku-go project.

Captain Satake, ordered by Lieutenant General Tada to put together a roster of researchers, went to the MOC Radio Technology Council for talent. The Council's recommendations led to the hiring of Dr Sasada Sukesaburo. In addition, Sone Tamotsu, an MOC researcher working on television technology, and four of his subordinates joined the effort. The Army also brought in top consultants from around Japan, including Drs Ikebe Tsuneto, Nishimaki Masao, and Okabe Kinjiro. Ikebe, a physicist with overseas experience in Great Britain and the United States, worked with Nishimaki, an assistant professor at Tokyo University of Engineering, on developing magnetron technology for the death ray. The MOC's Dr Kusunose oversaw the manufacturing of vacuum tubes, with the technician Sone Tamotsu his key subordinate. Okabe, a professor at Osaka Imperial University's Institute of Scientific and Industrial Research, was a world pioneer, at first under the direction of Dr Yagi Hidetsugu, in magnetron research. He was awarded the Imperial Prize of the Japan Academy in 1941for his work.[14]

The technician Yamada Genzo, involved in building transmitters, recalled years later that the facility was on the former site of the Japan Higher Colonisation School, which had been established to train Japanese emigrants to settle in Brazil's Amazon region. Noborito made use at first of the site's several buildings, which included a kendo training hall. In fields around the hill were planted peach trees. One of the guards, previously a local farmer, would bring seasonal fruits and vegetables to the facility. For Yamada, such an idyllic setting was 'a fitting environment for researching dream-like new weapons.'

The Army soon created new structures and installed new equipment for its projects. Yamada recalled the construction of two laboratories, each approximately 9 by 36 metres, and a building of the same size to house the supply of direct current. To power the projects, ASRI technician Katsuki Suesuke came in August 1938 to Noborito to oversee installation of equipment for a direct-current power supply capable of supplying 10 kilovolts (kV) of direct current for experiments. By comparison, Japanese residences today run on 100 volts of alternating current. Japanese companies contributed to the effort. Toshiba supplied six-phase full-wave rectifiers. NEC provided its TW-530-B, a water-cooled vacuum tube, at that time the largest short-wave vacuum tube in the world. The new equipment was installed by the end of that year.[15]

Research then began in earnest at the beginning of 1939. Already, in 1937 and 1938, researchers had been using an oscillator to direct three-metre waves at engines to disrupt them. If the tests proved successful, the Japanese Army would be on its way to developing a revolutionary anti-aircraft weapon. The researchers did in fact succeed against exposed engines. However, according to Satake, 'Electromagnetic shielding had been fully conducted on aircraft engines to eliminate interference from onboard radio equipment, so there was no longer any gaps for electric waves to exit or enter.' Wave technology that could have threatened the fighter planes of the First World War could not down modern aircraft. In 1940, the researchers halted their research on stopping engines.

Having reached a dead end on using electric waves against internal combustion engines, Noborito's researchers turned in 1940 to investigating how to use electric waves to harm or kill the enemy. At first, they tested the effects of an electric field on a caged laboratory animal placed between condenser plates, according to Satake, sacrificing in the process many mice that 'died in a short period of time.' The basic arrangement was to irradiate an animal at the focal point of an ellipsoidal reflector fed by a dipole antenna that was, in turn, energised by a high-power, continuous-wave oscillator. Having demonstrated that the waves killed, the researchers then moved to send waves over increasing distances. Noborito tested mice, rabbits, marmots, dogs, and monkeys.

Over time, Satake recalled, researchers were able to place the specimens at increasingly greater distance. At distances 'greater than 10 metres,' the animals would die 'in about a minute or two.' Satake wrote that compressing the wavelength and increasing the electric power became the decisive problems.' A postwar report shows progress over time. Noborito researchers in 1937 were experimenting with oscillators producing a wavelength of 3 metres and power of 1kW. In 1944, they were using magnetrons to produce waves as short as 0.4 metres with an output of 30kW to focus ultrashort waves at a distance of 30 metres.[16]

Examining the deceased laboratory animals, researchers noted that the waves would raise the body temperature. Sasaki Tadashi, a graduate of Kyoto Imperial University and a specialist in vacuum tubes, whom the Japanese Army had assigned to Noborito, recalled working on the death ray. He described the IJA's idea as one of developing a weapon to 'irradiate people with microwaves.' Sasaki would monitor body temperatures by attaching thermometres to the lab animals.[17] In effect, Noborito was a pioneer in technology that would lead after the war to the production of microwave ovens.

In October 1941, according to a postwar investigation, Japanese military authorities 'diverted' Noborito's 'initial group' of researchers to work on radar elsewhere before 'renewing' work on the death ray in 1943. However, another postwar report indicated that experiments continued without interruption in the period of 1941-1943. It may be that those sent to work on radar were experts in electronics, given their assignment to the similar field of radar research. One example was that of Satake. Assigned to Noborito in 1940, he transferred in 1943 to the Tama Army Technical Research Institute, where he worked on technology to counter the American B-29 bombers.[18] Research from late 1941 into 1943 likely shifted in part or entirely from some aspects of equipment development to more of a focus on animal experimentation.

Indeed, leadership changes in this period suggest that this was the case. In January 1938, Major Kusaba Sueki was chief of what was then the Noborito Laboratory, where he directed 30 men in ultra-shortwave research. The following January, he was still overseeing 30 researchers; Dr Sasada Sukesaburo, an expert on the physiological effects of electric waves, worked under Kusaba as a civilian technician.

By January 1940, Colonel Shinoda Ryo was in charge of the Noborito Branch Depot, assisted by Dr Sasada, Major Satake, and a Lieutenant Colonel Watanabe. In 1942, Shinoda was directing what became that year the 9th ATRI. Within First Section, Dr Sasada 'and ten men conducted research on ultra-shortwaves and balloon bombs.' It was in 1942, when research on the equipment seems to have stopped and a smaller group worked on the biology, that Dr Sasada's reduced group experimented with dogs and monkeys.

The Army 'actively' renewed in 1943 research on the death ray, according to a timetable that was part of the postwar US military report. Much of the work that year involved designing an oscillator with 'higher power.' By December 1944, Dr Sasada's team had increased to 15 men. Using a magnetron with 30kW output power and 80cm wavelength that powered a dipole feeding a reflector of one metre the researchers that year reported that they were able to kill rabbits at distance of 30 metres in 10 minutes. The Army appropriated what was then the impressive sum of a million yen for further research in 1945.The US Army report stated: 'The experiments planned in 1945 were to use four 300kW (input) magnetrons in parallel, which were expected to give a total output of 250 to 300kW. These were to feed a dipole in a 10-metre diameter ellipsoidal reflector.' If successful, this system would kill a rabbit in 10 minutes at a distance of one kilometre.

From its low point, when Dr Sasada led a small team of 10 men, Lieutenant General Kusaba came to direct 116 (20 technical officers, 4 civilian engineers, 12 consultants, and 80 technicians) on the project. Having earlier divided his energies in overseeing both the development of bombing balloons and the death ray, by 1945 he was devoted to the latter project alone.[19]

IJA Wonder Weapons, Covert Operations

Pushing the frontiers of science and technology became increasingly important as Japan's advances at the war's start slowed, stopped, then turned into retreat. The Japanese Army and Navy had won tremendous early victories from December 1941 into the spring of 1942, including the conquest of the Philippines; Guam; British Malaya, with its island bastion of Singapore; Hong Kong; Burma; and the Netherlands East Indies. However, ominous signs appeared with the Japanese failure to prevail at the Battle of the Coral Sea (May 1942), preventing the seizure of New Guinea's Port Moresby, which would have threatened Australia. The Japanese loss at the Battle of Midway (June 1942) ended the possibility of Japan invading or neutralising the US territory of Hawaii as a key forward staging area. The American Doolittle Raiders in April that year pierced Japan's air defences to bomb Tokyo, the imperial capital. Successive reverses, an important one of which was the retreat in early 1943 from Guadalcanal in the Solomon Islands northeast of Australia, suggested that the Japanese Army would need to field new weapons against foes whose combined industrial bases dwarfed that of Imperial Japan. As Noborito technician Yamada Genzo recalled, the IJA put 'massive' sums into the Ku-go project at Noborito with the aim of developing a 'secret weapon that would turn the tide of war at a stroke' in Japan's favour.[20]

Lieutenant General Tanaka Shinichi, chief of AGS First Bureau (Operations) and several subordinates composed a document, dated 15 August 1942, soon after the Battle of Midway, calling for the development within one or two years of 'decisive weapons for the successful conclusion of the world war.' These innovative arms were 'to render useless' the enemy's 'existing weapons'. Proposed areas of research and development included both 'weapons to defend strategic areas' and 'epoch-making improvement in electric-wave weapons'. Both proposals involved the development of a death ray.[21]

Noborito at first had been a site for research and development in electric-wave technology as an offshoot of the ASRI. However, Satake wrote after the war, the focus of Noborito's activities changed over time to clandestine operations and special weapons.[22] This happened relatively soon after the ASRI's test site began operating as the ASRI Noborito Laboratory in late 1937.

An increasingly urgent need to end the war in China was one reason for this. The Japanese Army had waged a number of offensives and taken Nanking, the Chinese capital, yet China had not surrendered. Instead, the Nationalist government decamped to Chungking in southwest China to continue the fight. Bogged down in China's vast territory, anxious not to weaken its position against the Soviet Union by diverting increasing numbers of troops to China, the Japanese Army looked to unconventional operations. In April 1939, a reorganisation put research on electric waves at Noborito under a new First Section and added Second Section for biological and chemical weapons. In August that year, the Army established Third Section for counterfeiting and forgery. In a reorganisation, the site became in September the ASRI Noborito Branch Depot. In 1943, when the Japanese Army decided to develop bombing balloons, First Section was no longer working on electric waves alone.[23]

A Demonstration for a Prince

Under the pressure of war, the Imperial Japanese Navy (IJN) also started working on a death ray. Following the defeat at Midway in June 1942, according to an IJN technician's account, Admiral Yamamoto Isoroku called for the development of 'epoch-making weapons' to turn the tide. The commander of the Combined Fleet said at a technical conference that he would be unable to last another year by conventional arms alone. The next day, an officer of the Navy General Staff asked Lieutenant Ito Yoji, an ordnance officer, if there were any technologies with such potential. Ito suggested atomic power and a microwave death ray.[24]

The IJN had been investigating how to turn electric waves into weapons since the previous decade. Ito had written six reports in the early 1930s on using short waves in naval applications. When the Navy heard that Noborito had succeeded in 'transmission of a three-metre wave with 50kW,' as Noborito technician Yamada Genzo recalled years later, Navy Captain Prince Takamatsu, a younger brother of Emperor Hirohito, arrived at Noborito with approximately 20 technical officers early in 1940 for a demonstration. A reviewing stand was placed 1.5 metres in front of a vertical dipole antenna to demonstrate how several seconds of exposure to high-frequency waves would warm a human body. The uniformed visitors were cautioned to remove their swords during the demonstration. However, the officer from Prince Takamatsu's group chosen to mount the reviewing stand forgot to leave behind his sword. The moment that the demonstration began, the officer bellowed and leapt from the reviewing stand from the electric shock he had received.[25]

Serious Doubts, But Great Potential

In 1935, Robert Watson-Watt, superintendent of the Radio Department in Great Britain's National Physical Laboratory, reported to the

Aeronautical Research Committee under Sir Henry Tizard that the death ray was an impossible idea. He suggested instead the application of wave research to radio direction and ranging as a way to spot enemy aircraft at great distance. As for a death ray, Watson-Watt's assistant, Arnold Wilkins, calculated that such a weapon was impractical to build with the technology then available. Using radio waves to detect distant objects was possible.

In the United States, too, there was little enthusiasm for developing a death ray. Stanley Lovell, a member of the National Defense Research Council before becoming director of OSS/R&D, dismissed the idea in a postwar memoir. He described his response to a hypothetical wartime mission given him on how, going ashore at night on a coast under German occupation, he would destroy a guarded wireless installation: 'I early abandoned such fantasies as a death ray, which I knew would require a great power plant to implement it.'

Doubting the practicality of a death ray, both the British and the Americans worked to apply electric waves to detecting enemy aircraft and ships with the emerging technology of radar. They succeeded in developing and deploying functioning radar systems in the Second World War. Why, then, did the Japanese Army devote substantial resources to the death ray? Why not devote all efforts on electric waves to developing radar? After all, Noborito was also a site for radar research before the Army consolidated research in that area of research at the Tama Army Technical Research Institute. Noborito technician Yamada Genzo suggested that Japanese military officers responsible for commanding troops in battle, who had greater influence than technical officers, considered communications and electrical equipment to be 'defensive' and, therefore, 'secondary.' A postwar historian of Noborito argued that the IJA decision to favour 'the more offensive' was characteristic of IJA military thinking.[26]

Seeking to conduct cutting-edge research and development under wartime conditions must have put tremendous pressure on Noborito personnel. Technical Major Yamada Genzo after the war described the two goals of using electric waves to stop engines and kill the enemy as 'fantastical research topics.' Yes, Noborito researchers had succeeded in killing laboratory animals at some distance with electric waves, but Yamada 'had great doubts', given the huge amount of energy required, whether Noborito would be able to produce a weapon capable of killing enemy forces out in the open.[27]

For Colonel Satake Kinji, the story of Noborito from beginning to end was one of research on ultrashort waves. The Army's vision of a death ray was, perhaps, more fantastical than practical. Still, as Satake wrote, such was the nature of research. 'If we think about it, it was a huge waste. However, waste accumulates and becomes a large foundation. I think that there is probably more waste than success in this world.' As for turning electric waves into an offensive weapon, Satake recalled early doubts about media reporting in the 1920s on electric waves stopping a car engine in Germany. On the other hand, if true, then its 'effect would be outstanding'. Such thinking led, according to Satake, to 'one investigation after another'.[28] It led to Noborito.

CHAPTER 3

PREPARING BIOLOGICAL WEAPONS

Veterinary science and veterinary scientists exist to promote the welfare of the people. But the livestock hygiene institute where I worked in the days of Japanese imperialist rule was engaged in scientific research in top secret, not for the people but against them.

– Kim Jong Hui, technician, Livestock Hygiene Research Institute, Office of the Governor-General of Chosen[1]

Biological Operations in the Shadows

Killing with poisons or pathogens while hiding one's hand has a long history. In Renaissance Italy, the Borgia and Medici families were notorious for eliminating rivals by secretly poisoning them. In the First World War, Germany conducted secret biological operations against livestock. Governments continued after that war to produce biological weapons for covert and special operations, including pens with concealed poison pellets of the Soviet Committee for State Security (KGB).[2]

Such clandestine operations are related to biological warfare, but tend to be smaller in scale and covert in nature. The German agents who stealthily infected American livestock bound for Allied ports in the First World War executed clandestine biological operations. Nations engaging in biological and chemical operations and warfare have had different organisations for developing the weapons involved. The US Army's Camp Detrick in Maryland, for example, conducted 'conventional' BW research for large operations. The OSS in the Second World War conceived a plan to covertly infect individual Japanese military officers with botulism.[3]

In Japan, the Imperial Japanese Army had different organisations for overt and covert biological warfare. The IJA Kwantung Army Epidemic Prevention and Water Supply Department, designated Unit 731, was the main unit of a BW network whose field locations extended from Manchuria to Southeast Asia. Another notable unit was the Kwantung Army's Military Horse Epidemic Prevention Depot (Unit 100), which was responsible for developing weapons against horses and other animals. In China, notable IJA organisations were Unit 1644 (also known as Unit Ei 1644) in Nanking and Unit 8604 (also known as Unit Nami 8604) in Canton.[4]

For secret biological operations, the Japanese Army turned to the 9th ATRI at Noborito. Of the IJA's ten numbered laboratories, only Noborito operated under AGS Second Bureau's 8th Section (Clandestine Operations). Noborito's Second Section carried out research and development for biological warfare. Of the section's seven groups, Groups 2 and 3 handled poisons; Groups 4 and 7 developed materials to kill animals, and Group 6 worked on weapons to destroy crops and other plants. Even Group 1, which made special equipment under the direction of Technical Major Ban Shigeo, was involved with biological and chemical research and development.[5]

Poisons for Covert Operations

Second Section's Group 3 worked on poisons from various sources, according to Ban. Those sources included plants, which yielded aconite and other poisons; snakes, including cobras, for their venom; marine animals, such as puffer fish, which produced tetrodotoxin; inorganic materials, including thallium; and organic substances, such as phosgene. One member of Group 3 recalled accompanying his group's director, Technical Major Hijikata Hiroshi, to Formosa to acquire venom from habu, a local species of pit viper, to put into fountain pens. Researchers divided their work into two categories: poisons that worked immediately on the victim and those with a delayed effect. The latter could help hide a poisoner's hand by putting time between the target receiving the poison and dying from it. Ban wrote long after the war that Group 3 also sought to develop new poison compounds without colour, odour or taste. The group's members developed such poisons and concealed them in sweets, chocolate, coffee, fruit, medicine, and other products.

Major Hijikata was an expert in pharmacology who had graduated from Meiji Pharmaceutical School. His principal researcher was Captain Takiwaki Shigenobu, who had arrived at Noborito in 1939 from ASRI Department 3. After attending Chiba Medical School, Takiwaki had worked at the Tokyo Hygienic Laboratory. Years later, he told a writer gathering material on Noborito that the IJA had ordered him there to develop 'practical weapons using hydrocyanic acid'. Takiwaki developed at Noborito a substance derived from acetone cyanohydrin, produced largely from liquid hydrocyanic acid supplied by the 6th ATRI. The poison had a delayed effect. Stable when placed in an ampoule, making it easy to store and transport, the product was excellent as a poison for clandestine operations. Takiwaki recalled that Noborito produced approximately 500 ampoules with a volume of one milliliter. He claimed to have heard that the Nakano School took delivery of them via the Army General Staff and that unused ampoules later came back to Noborito's Second Section for storage.[6]

As with other Noborito spy gear and special weapons, officers of the Kempeitai and the Nakano School used poisons abroad against foreign targets of assassination. However, one of the few surviving documents from Noborito suggests that the Japanese Army also possibly used Noborito poisons for assassinations in Japan. In the document in question, dated 8 February 1943, Ban sent Lieutenant Colonel Soda Mineichi, a Kempeitai officer in the War Ministry's Defence Section, which administered the Yama Agency, a report on drug tests. A researcher on Noborito's history has written that Ban at this time had recently completed testing on various poisons. She has also cited a history of the Nakano School produced by veterans that identified Soda as a Yama officer. Therefore, according to the researcher, 'it can be inferred' that the Yama Agency was carrying out assassinations in Japan or at least was interested in doing so.[7]

Award and Atonement

Ban finished the testing some time around February 1943, according to the document. In April that year, the War Ministry awarded him a certificate of technical merit, dated 14 April, bearing his name and that of Noborito's director, Shinoda. The document was in the name of Army Minister Tojo Hideki, the general then leading Japan in his concurrent post of prime minister. The award, for the development of 'special physical and chemical scientific materials', a reference to the acetone cyanohydrin and other materials that Ban and his colleagues had been developing, came with a prize of 10,000 yen, worth roughly 100 million yen today.

The recognition and award money must have gratified and troubled Ban and Shinoda. It was a high honour to receive a certificate of merit from Prime Minister Tojo for their work. Yet, they must have felt distressed by what they had done to develop the prize-winning poisons. Soon after receiving the award, Shinoda erected at Noborito an imposing stone monument to appease the souls of animals sacrificed in Noborito projects. It was a monolith approximately 2.7 metres high, 1 metre wide, and 15 centimetres thick. Shinoda surely must have raised such a grand monument for some reason beyond the many mice and rabbits killed by electric waves at Noborito. Shinoda also used the award money to build at the institute a Shinto shrine, the Yagokoro Shrine, in part to hold memorial services for personnel who died in laboratory accidents.[8]

Gathering Materials, Experimenting on Prisoners in China

Noborito's work in China was what must have troubled Shinoda and Ban. In May 1941, Colonel Shinoda had ordered Lieutenant Colonel Hatao Masao, then director of Second Section, to lead a team to Nanking. Ban, who was chief of Group 1; Hijikata, head of Group 3; and several subordinates accompanied Hatao. Shinoda had coordinated the visit with Major General Ishii Shiro, an IJA medical officer who was the leading force behind the military's BW work in Manchuria and China as director of Unit 731. The visitors from Noborito were to participate in the testing on prisoners of cyanide poisons, which had already proven lethal in animal experiments. Unit 731, which was primarily working in Manchukuo to develop pathogens, also had a pharmacological section. The IJA Central China Expeditionary Army Epidemic Prevention and Water Supply Department (Unit Ei 1644) would supply the doctors. The test site would be a military hospital formerly under control of the Chinese Nationalists. Chinese, prisoners of war and criminals already sentenced to death, would be the test subjects.[9]

Results varied according to the method of delivery for the poisons, primarily acetone cyanohydrin, according to one witness. Injected prisoners died within two or three minutes. Those who drank the poison were dead in approximately five to 10 minutes, according to one of Hijikata's subordinates. The rapid deaths suggested that those prisoners had died from cyanide poison. Ban recalled prisoners taking longer to die with Noborito's poison. He wrote that the effects of acetone cyanohydrin were apparent in two or three minutes, with death generally a half hour after ingestion. Depending on age and other physical factors, however, death could take as many as 10 hours. Beyond hydrocyanic acid, according to another account, the Noborito group also brought with them snake venom, other poisons, and even strains of bacteria for testing on prisoners.[10]

The Noborito Research Institute appears to have engaged repeatedly in joint work and exchanges of data with Unit 731 and other IJA units. Koshi Sadao, personal driver to Unit 731's director, stated at a 1989 history symposium that Noborito had engaged in exchanges of data and materials with the Army Medical School and with Unit 731, which acted as a field unit performing experiments for the school. Koshi claimed that Unit 731 had the facilities and the 'materials,' an apparent reference to prisoners for testing, that were lacking at Noborito.[11]

Noborito also worked with Unit 1644 and other BW units in Asia to test their products outside Japan. Researchers of Second Section's Group 6, who were developing diseases and methods to devastate crops in China and the United States, avoided the danger of harming Japanese harvests by conducting dangerous field tests outside Japan. They worked with IJA units overseas to conduct their experiments against foreign crops. Group 6 technician Matsukawa Hitoshi relied in his 1942 experiments in central China on Japanese military aircraft to release rice borers and fungi causing stem rot over Chinese rice fields west of Tungting Lake to ascertain their effectiveness against crops by aerial dissemination.[12]

Although weapons for the battlefield and those for clandestine operations are in some cases developed in separate organisations for different purposes, the experiments in Nanking by members of Noborito and Unit Ei 1644 were joint. Unit 731 in Manchukuo and affiliated military units in China and elsewhere, including Unit 1644, indeed were developing biological weapons for the battlefield, but personnel there worked with Noborito on weapons for clandestine operations.

A prominent example of someone who worked in both battlefield and clandestine applications was Lieutenant Colonel Naito Ryoichi, a medical officer at the Army Medical School's Epidemic Prevention Research Laboratory in Tokyo. Naito worked with Unit 731 and the 9th ATRI, exchanging data with Ishii's officers in Manchukuo and working with Shinoda's men at Noborito. He was a participant in the Fu-go project, an indicator of the IJA's intent to use the bombing balloons to carry biological weapons. According to a postwar expert on the history of Noborito, Naito's joining Fu-go also suggests that the balloon project brought together resources from Noborito, the Army Medical School, and Unit 731.[13]

Moreover, the Japanese Army carried out both conventional and clandestine operations in military campaigns. In the summer of 1940, IJA aircraft released plague-infected fleas from Unit 731 over Chinese positions in the area of Ningpo, Chekiang Province, as part of a military campaign. In 1942 approximately 100 men of Unit 731 joined the Japanese military campaign in Chekiang and Kiangsi provinces. As Japanese forces withdrew at the campaign's end, members of Units 731 and Ei 1644, disguised in Chinese clothing, placed Chinese sweets and other food laced with pathogens in various places to infect the local population. In Okinawa, late in the war, the Japanese Army waged a tenacious defence with conventional weapons. In addition, military operatives in plain clothes arrived before the fighting with pens from Noborito to release bacteria into the water supplies of areas overrun by the invading Americans.[14]

The Poisoning of Li Shih-chun

A slow poison is an ideal assassin's tool. The poisoner has time to escape the scene, perhaps even avoid coming under suspicion for the killing. The circle of suspects widens as time passes, obscuring or hiding the hand of the assassin. Imagine a victim ingesting without suspicion such a poison with dinner at a restaurant, then proceeding to a night club or two before returning home to die in bed in the early morning hours, or perhaps at work later. The longer the lapse in time, the less clear the trail of evidence. The poisoner escapes.

The Japanese Army, mired since 1937 in costly and unending military conflict in China at a time when the Soviet Union posed the foremost threat to Japan's position in Asia, looked for ways to stop the fighting. Following Prime Minister Prince Konoe Fumimaro's 1938 declaration that Tokyo would no long deal with Chiang Kai-shek's Nationalist government, the IJA made various discrete approaches to open backdoor negotiations with Chungking.[15] The Japanese Army, via Second Bureau's 8th Section, also put in place in 1940 in Nanking a rival Nationalist regime under the prominent defector Wang Ching-wei. A third option was to kill Chiang.

Colonel Okada Yoshimasa recalled years after the war how the Japanese Army had considered assassinating the Nationalist leader. Okada, an infantry officer from IMA Class 36 (1924), was experienced in Chinese affairs. His assignments included a posting as a staff officer in 1939 to the China Expeditionary Army General Headquarters in Nanking, a concurrent assignment in 1942 to the Shanghai Army Bureau and a subsequent posting as a senior staff officer to Twenty-Third Army Headquarters in Canton. In 1938, assigned to AGS Second Bureau's 8th Section, Okada met in Tokyo at the Imperial Hotel for a week of discussions with the agent Sakata Shigemori, an old associate of his with a long history in China. The two concluded that the best solution for ending Chinese resistance was to assassinate Chiang. As an alternative, the two men favoured wrecking China's economy by flooding the country with counterfeit Chinese currency.[16] In the end, as we shall see in a later chapter, the Japanese Army chose the latter option.

Colonel Okada did not explain how he and Sakata had considered killing China's leader, but poison appears to have been considered. In 1998, Japanese media reported the surfacing of notes from a deceased officer of the Tokyo Metropolitan Police Department (MPD) from an infamous poisoning case of 1948: the Teikoku Bank Incident. In the course of investigations, the MPD had questioned a wide circle of suspects, focusing on military veterans. The MPD officer's notes indicated that the Japanese Army had drafted a plan to poison Chiang Kai-shek. The media further reported that the notes indicated that Noborito had developed poison for such an operation. Under police interrogation, former IJA technicians had spoken of human experiments conducted in the years 1941-1944 in Nanking and Shanghai. According to the notes of their testimony, Japanese technicians had given 'thousands' of Chinese prisoners such items as tea, milk, and whiskey laced with lethal substances, or simply injected doses and noted the effects.[17] Major Ban's 1941 work as a member of a Noborito delegation in Nanking, including cooperation with members of Unit 1644 on human subjects, appears to have been part of the activities to which the technicians confessed.

The Japanese Army did not assassinate Chiang Kai-shek but did remove a lesser thorn in its side by poisoning a key security official in the Nanking government. Wang Ching-wei, a senior Nationalist official, had left Chunking in 1938 before assuming leadership of a rival government in Nanking in 1940. Without his own military or security forces, Wang depended on the Japanese Army for security. He also had a local security service. Wang's subordinate, Vice President Chou Fo-hai, chaired Nanking's Security Service Committee and the Committee for the Elimination of Counterrevolutionaries. Ting Mo-tsun ran the Security Service General Headquarters (SSGHQ) with Li Shih-chun the second in command. The SSGHQ operational headquarters, backed by the Japanese, was located in Shanghai at No. 76 Jessfield Road. Ting and Li worked in the shadows from that address to cajole, cow or kill enemies of the Nanking regime.

In 1943, the Japanese Army turned on an increasingly ambitious Li. He had by then alienated SSGHQ Director Ting as well as Vice President Chou. Worse, Li had lost due to transfers out of the area the previous year his chief Japanese protectors: Major General Kagesa Sadaaki, in charge of the Nanking regime's security as its military advisor and chief of the IJA Ume Agency, and Lieutenant Colonel Haruke Yoshitane. The new chief of the Ume Agency joined his subordinates and officers of the Kempeitai in deciding eliminate Li. The task fell to Lieutenant Colonel Okamura Tekizo of the Kempeitai. On 6 September, Okamura served Li a fatal dinner. The meal came with several dishes, but Okamura insisted that he try the meat pie. Poisoned, Li fell ill the next day and died two days later.[18]

Did the poison come from Noborito? Developing poison without colour, odour or taste to conceal in food and drink was a major activity for Second Section. The delay between Li's dinner and his death, consistent with Noborito's work on poisons with delayed effect, also suggests Noborito's work. On the other hand, Lieutenant Colonel Otsuka Kiyoshi, a Kempeitai officer who had worked with Okamura in China, told an SSGHQ veteran many years after the war that Unit 1644 had supplied the poison.[19] Given Noborito's responsibility for developing tools of assassination and the joint work in China between Noborito and Unit 1644, however, one can well imagine that Unit 1644's supplies included poisons from Noborito.

Cattle Plague on the Peninsula

Members of Second Section's Group 7 worked on weapons to harm or kill animals. Noborito built facilities for animal experiments, with an emphasis on cattle, horses, and pigs. As with researchers in the groups responsible for developing poisons and pathogens to use against people, those in Group 7 were part of a larger IJA effort to develop biological weapons.[20]

When Colonel Shinoda had in 1940 the idea of building a laboratory at Noborito for work against cattle, horses, and hogs, according to subordinate Ban Shigeo, the Army Medical School supplied pathogens. Ban described the division of labour: Unit 731's Ishii took the lead in developing biological weapons against people, Shinoda against animals.[21] IJA Unit 100, too, worked on animal projects. If there was a division of labour, perhaps it had to do with scale, with Unit 100 developing weapons for military campaigns and Noborito developing tools for special operations.

Kawashima Hideo came to Noborito in 1940 as a consultant to help realise Shinoda's plan to turn animal diseases into weapons. Dr Kawashima, an authority in bacteriology and virology, had read microbiology in the Veterinary Department of Tókyo Imperial University's Faculty of Agriculture before becoming a technical expert and senior researcher in the bacteria section of the Animal Infectious Disease Research Laboratory of the Ministry of Agriculture and Forestry (MAF). In addition to writing multiple reports for the MAF laboratory, he authored in 1944 a book on cattle diseases.[22]

In addition to consulting for Noborito, Kawashima also scouted talent. On a recruiting trip to the Army Veterinary School early in 1943, he interviewed a student, Kuba Noboru. Prior to his studies at the Army Veterinary School, Kuba had served in China in with the Shanghai Expeditionary Army's 2nd Field Chemical Laboratory. The young veterinary student evidently impressed Kawashima. Kuba, for his part, must have heard something in Kawashima's description of Noborito. Following the interview and the recommendation of one of the school's instructors, Kuba joined Group 7 of Noborito's Second Section on graduating the Army Veterinary School that year. He would became chief of the group and reach the rank of major.

Kuba took the train each morning to Noborito, walking from the station up the hill to work, then retracing his path at the end of the day.

As he walked, Kuba would think of how to wreck the enemy's will to fight. Infecting American's cattle herds with rinderpest, also known as cattle plague, struck him as a solution. To make the disease effective as a weapon would require heightening the virulence of the viruses, freeze-drying them, and storing them in powder form.[23] The project would also require cattle and land for experiments.

For his project, Kuba tapped existing resources at Noborito and developed new ones. His mentor Kawashima, working at Noborito on contract, was already available for consultation. Kuba looked across the Sea of Japan to the Livestock Hygiene Research Institute, which operated under the Office of the Governor-General of Chosen (OGGC). Located in the southern port city of Fusan, known after the war as Pusan, the institute produced vaccines for cattle and other livestock. In 1911, the year after the incorporation of Chosen (Korea), into the Japanese Empire, the Agriculture Ministry had established in Fusan the Rinderpest Serum Plant to prevent the disease's spread on the peninsula and into the Japanese home islands. Administration was later transferred to the OGGC, the plant's mission was expanded to cover other animal diseases, and it was renamed in 1918 the Veterinary Serum Plant. It became in 1942 the OGGC Livestock Hygiene Research Institute.[24]

The leaders of the Livestock Hygiene Research Institute were top-tier veterinarian scientists, graduates of the Department of Veterinary Medicine at Tokyo Imperial University. Isayama Isaburo directed the institute. A 1915 graduate, Isayama served as a civilian senior technician for the Japanese Army in China before working in the OGGC from 1925 and rising to become director of the Livestock Hygiene Research Institute in 1940. Nakamura Junji, who oversaw production, had joined the institute as a technician upon graduating from the Department of Veterinary Medicine in 1926. Ochi Yuichi, who directed research, Nakamura's university classmate, had also gone after graduation to the Livestock Hygiene Research Institute.[25]

In 1944, Kuba gained the cooperation of Isayama and Nakamura, enlisting the two to work on contract with Noborito. Kuba also brought in Hotta Tokuro, a graduate of Tokyo Higher Veterinary School who was working in the OGGC Livestock Hygiene Research Institute's Vaccine Department, and Sasaki Kiyotsuna, an expert on Soviet livestock and a professor in the Livestock Department of Tokyo Imperial University's Faculty of Agriculture. For administration, Kuba hired Iwasa Takeshi.

Together with Kawashima and Nakamura, Kuba drafted the following goals to guide the project:

a) explosive spread of cattle plague and hog cholera;
b) priority on cattle plague, followed by hog cholera;
c) isolation of the highly virulent rinderpest virus found naturally in Manchuria;
d) increasing the toxicity by culturing the isolated virus in cattle;
e) freeze-drying of the virus and its rendering into a powder to preserve virulence;
f) conducting of field tests on cattle using the highly virulent powdered virus;
g) annihilation of cattle using bombing balloons carrying the highly virulent powdered virus.[26]

Producing a freeze-dried and powdered rinderpest virus would enable the Japanese Army to overcome the severe environment that the bombing balloons would traverse in carrying their biological payloads to the United States. The balloons were to soar thousands of metres into the sky, where temperatures would drop to 50 degrees below zero. Unit 731 and other IJA units had worked to drop plague-infected fleas and infected materials from aircraft flying at relatively low altitudes. Such insects and materials would perish on the balloon trips across the Pacific Ocean. Kuba's project would produce a viable payload for the balloon attacks.

Nakamura was the project lead in Fusan. Somewhat unsocial, he had chosen to study veterinary medicine because he thought it would suit his nature to treat animals rather than people. He had applied to work at the serum plant because of its good reputation, due in part to the practice of Kasai Katsuya from Tokyo's prestigious Kitasato Institute and other prominent scientists going to Fusan in the summer to guide research and production at the institute. Nakamura had built a successful record over time as a veterinary scientist. In a period of 15 years, he succeeded in creating an attenuated vaccine for cattle by passing it through rabbits (a lapinised vaccine). His vaccine proved even more effective than that of Kakizaki Chiharu, who had earlier produced in Fusan a rinderpest vaccine. Nakamura had obtained a doctoral degree in agriculture in 1941 and was directing production at the Livestock Hygiene Research Institute when Noborito enlisted him.[27]

Isayama and Nakamaura, for their part, almost certainly brought one or more members of the local Korean staff into the project. One of them was Nakamura's 'favourite apprentice,' Kim Jong Hui. Born in 1909 into a relatively well-off family in South Heian Province, Kim attended a higher normal school in the colonial capital of Keijo. As was the case with many ambitious and talented young Koreans, Kim left the peninsula for the Japanese metropole. He studied veterinary and livestock science in the Hokkaido Imperial University's Faculty of Agriculture. Failing to find satisfactory employment in Japan despite having graduated with high marks in 1938 from an imperial university, a failure he blamed in his postwar memoir on Japanese discrimination against him as a Korean, Kim returned to the peninsula to work from 1942 in Fusan at the OGGC Livestock Hygiene Research Institute.[28]

Kim gave no specifics on his position or activities in Fusan, but he appears to have been working closely under Nakamura Junji, who was directing production. Therefore, Kim was likely working to produce vaccines for rinderpest and, perhaps, working under Nakamura to develop rinderpest as a weapon. Kim's contention that the Fusan institute was 'engaged in scientific research in top secret, not for the people but against them,' seems almost certainly to be a reference to Noborito's rinderpest project. He further asserted: 'After I saw this, I began to feel the bitterness of disillusion about veterinary scientific pursuits.' In light of his statements and his close relationship with Nakamura, Kim seems at a minimum to have known of one of Noborito's most secret projects.[29]

Progress in making a weapon of rinderpest came with samples gathered by Hotta in April 1944 from cattle that had died from it in Tongliao, a site in Inner Mongolia near the border with Manchukuo. In Fusan, Hotta and others isolated the virus, enhanced its virulence, dried it to make a powdered pathogen, then froze it. The team's powdered virus retained its virulence even when exposed to light and freezing temperatures. In May, Kuba arrived in Fusan from Noborito at the head of a six-member group to join in field tests. Kuba, Nakamura, Hotta, and the other participants witnessed the dispersal of the freeze-dried powdered virus by a small explosive charge over a field where 10 head of cattle were staked. Within three days, all the animals had developed a fever. A week after their exposure, all the cattle had died.

Noborito, with Kuba as the point man, had succeeded in leveraging the resources of the Livestock Hygiene Research Institute to develop a biological weapon suitable for loading on bombing balloons and attacking the United States. Kuba years later wrote his appreciation for the institute's success in developing virulent rinderpest to 'attack enemy nations and annihilate them.' Kuba wrote that success in turning cattle plague into a weapon was due to 'the cooperation of OGGC Livestock Hygiene Research Institute Director Isayama and to the guidance of Dr Nakamura, as well as to his devoted efforts and those of Dr Hotta and the others involved.'

In September 1944, according to Kuba, senior officers assembled at the headquarters of the Army General Staff to decide whether or not to approve a plan to produce 20 tons of powdered rinderpest virus and load it on bombing balloons to attack the United States. Joining AGS staff officers at the meeting were Colonel Kusaba Sueki, director of Noborito's First Section, responsible for developing the bombing balloons; Major Kuba, representing Second Section's Group 7; Doctor Nakamura from Fusan; Lieutenant Colonel Wakamatsu of Unit 100; Lieutenant Colonel Kuchii Tadao, an officer of the Army Veterinary School; and Dr Nakamura Norichika, director of the MAF Animal Infectious Disease Research Laboratory. The conference participants agreed that the concept was technically feasible. However, General Tojo Hideki – still influential even after having resigned in July that year as Japanese prime minister, army minister and chief of the Army General Staff – reportedly rejected the plan due to concern that the United States would retaliate for losing cattle by attacking Japan's rice fields at harvest time.[30]

CHAPTER 4

BOMBING AMERICA BY BALLOON

> The bombing balloons – riding the winter jet stream at high altitude, flying 10,000 kilometres across the Pacific Ocean and attacking the American mainland – were truly dream-like weapons.
>
> – Major General Kusaba Sueki, director of First Section, Noborito Research Institute[1]

The Doolittle Raid of April 1942 on Tokyo and other cities shocked the Japanese public and stoked a keen desire among military leaders to strike the United States. US Army Lieutenant Colonel James Doolittle had led American pilots flying bombers from aircraft carriers against Japanese targets before proceeding to China. A similar Japanese counterstrike seemed beyond the range of possibility. The IJN defeat at the Battle of Midway in June 1942 threw Japan on the defensive. Sending Japanese carriers across the Pacific to launch air strikes against targets in California or elsewhere on the West Coast was scarcely possible. IJA airfields were far from the American mainland. No bombers could fly that far. In the face of such problems, Japanese military leaders considered both conventional and unconventional solutions.

Intercontinental Bombers and Bombing Balloons

The Japanese Army and Navy tasked the Nakajima Aircraft Company in 1942 in Project Z to develop bombers capable of flying from Japanese airfields to strike targets in the continental United States and continue to airfields in German-occupied France. From there the pilots would sortie for more bombing on the way back to Japan. Major General Sato Kenryo, director of the War Ministry's Military Affairs Bureau, and his IJN counterpart, Rear Admiral Ishikawa Shingo, burned with enthusiasm for taking the fight to the United States. The two had spoken of together flying a bomber into a New York skyscraper. As early victories gave way to defeats, however, the Japanese Navy lost enthusiasm for the project. Sato at one point made a passionate plea before a gathering in the auditorium of the War Ministry to strike the American mainland, but he could not save the bomber. Project Z ended in the summer of 1944 without producing the envisioned aircraft.[2]

In 1943, Japanese military leaders decided to strike the continental United States with bombing balloons. They chose such an unconventional approach because Project Z was far from producing new aircraft and neither the Japanese Army or the Navy had conventional means of bombing the American mainland. In August, the AOAH ordered the development of the balloons.

The Japanese Army by that time had been working for 10 years on special balloons. Major General Tada Reikichi had started the programme in 1933 as a department director in ASRI. The goal at that time was to launch balloons with a range of 100 kilometres from Manchukuo to bomb Vladivostok and other targets in the Soviet Far East. In line with the practice of designating projects with a Japanese kana character, activities involving these balloons (風船, *fu sen*, in Japanese) received the code name Fu-go. Following the development of balloons with a diameter of four metres, constructed of Japanese paper and paste made from the root vegetable konjak, the Japanese Army in 1939 organised a unit for the balloons as part of a meteorological regiment stationed in Manchukuo. In addition to the goal of dropping explosive and incendiary bombs, the ASRI also investigated using the balloons to scatter propaganda leaflets and, as an alternative to aircraft, to drop troops behind enemy lines.

The AOAH order of August 1943 directed the Noborito Research Institute to research and develop new bombing balloons with longer ranges. Required were balloons that would ride the jet stream at an altitude of approximately 10 kilometres over a distance of some 8,000 kilometres to reach North America. In the 13th century, storms had struck Mongol invasion fleets, seemingly heaven's intervention to help the Japanese samurai to stave off invasion. The Japanese had since viewed the timely storms as 'divine wind' (神風, *kamikaze*). In the modern empire's hour of peril, IJA officers hoped that another divine wind, in this case the jet stream delivering bombs by balloon, would save Japan from defeat.[3]

Colonel Kusaba Sueki, chief of Noborito's First Section, took overall responsibility for the project. His key subordinates included

Major Otsuki Toshiro, the lead for research on Type A balloons, made of paper; Major Takeda Teruhiko, responsible for working with the IJN to develop Type B balloons, made of rubberised silk, and Major Takano Yasuaki, who would work to develop improved radiosondes for tracking the balloons across the Pacific Ocean. Consulting for First Section were several eminent scientists, including Dr Yagi Hidetsugu. From Second Section, Major Ban Shigeo contributed by research on incendiary bombs. From Third Section, Major Ito Kakutaro worked to develop traditional Japanese paper that would not tear under pressure while the balloons flew across the ocean. The director of Fourth Section, Colonel Hatao Masao, was responsible for coordinating production of the new bombing balloons.

From beyond Noborito came participants from the army, the navy, universities, and industry. Under the AOAH, the 5th and 8th ATRIs worked to solve various technical issues, as did the Army Explosives Research Institute of the Second Army Arsenal. The Central Meteorological Observatory, working under the Army Meteorological Department, tackled meteorological issues. The CMO's Dr Arakawa Hidetoshi contributed greatly by charting the jet stream. From the Army Medical School, the IJA centre for biological warfare, came Lieutenant Colonel Naito Ryoichi. The Army Munitions Main Depot handled the supply of materials. The IJN contributed two lieutenant commanders for joint research. Tokyo Imperial University's Aeronautical Research Institute was also on board. From the private sector, Tomoegawa Paper and several other companies produced paper for the balloons. Several other companies, including Seikosha, manufactured parts.[4]

In the winter of 1943, Major Takeda took command at Ichinomiya, near Tokyo on Chiba Prefecture's coast, of a test group of approximately 100 members. The group launched approximately 100 test balloons and tracked them by radiosonde as far as Hawaii. Through repeated testing, Noborito developed what would be Japan's main bombing balloon, a spherical paper balloon 10 metres in diameter, to carry a payload of weapons across the Pacific Ocean.[5]

The balloon was made of traditional Japanese paper, made from the pulp of mulberry bark, laminated in several plies of paper and glued together with konjak, used in Japan for making noodles and other foods. An automated system kept the bombing balloons in flight. When a balloon dropped below a certain level, barometric switches would close, detonating a plug and releasing ballast – a sandbag – to maintain altitude.

Each balloon would carry three dozen sandbags on a circular carriage. Once the sandbags were all released in a period of time calculated as sufficient for the jet stream to have carried the object to the US mainland, the bombing balloon would begin dropping its explosives and incendiary devices. Following release of the last weapon, a demolition charge would blast apart the equipment. On some balloons, a charge would set fire to the bag itself.[6]

Noborito had developed a balloon that would float silently, perhaps undetected, to the United States, cause destruction and set fires, then self-destruct, leaving little trace.[7] Kusaba wrote in colourful language after the war of the weapon's desired effect:

> 'Today a fire occurs in Montana. The next day, several people die in an explosion outside of Los Angeles. Burnt scraps of what appears to be balloon paper are discovered at the foot of a mountain over there. An aluminium frame with sandbags attached to it goes rolling down a valley here. If such things happen again and again, rumors will spread, and there will certainly be at least a psychological effect. This was the aim of these weapons.'[8]

Writing in 1961 for a popular magazine, Kusaba omitted mention of the work to develop cattle plague and other biological payloads for the balloons. He did describe the plan to drop bombs from mysterious balloons to spread fear among Americans. Washington would have to devote resources to defend the nation against the silent paper bombers. Japanese authorities would be able to inform the Japanese public of successful strikes against the American mainland.

Manufacturing Bombing Balloons

Once Noborito had developed the bombing balloons, the Japanese Army transferred the programme in its production phase to the War Ministry and its AOAH. Lieutenant Colonel Kunitake Teruhito of the War Ministry's Military Affairs Bureau and Colonel Saito Tamotsu of the AOAH performed key roles in this phase. The Army budgeted 200 million yen for the production of approximately 20,000 bombing balloons. Major General Sato Kenryo, director of the Military Affairs

Bureau, had earlier ordered 250,000 of them, but Japan's resource constraints limited production to roughly a tenth of his desired number.[9]

Manufacturing the balloon bombs required a work force, work sites, and materials. The government had earlier arranged for workers to boost wartime production by issuing in August 1944 the Women's Volunteer Labour Ordinance to send young women to military production sites. After the ordinance went into effect, several tens of thousands of students, shop clerks, and other young women reported for duty at designated locations. Many of them worked to make bombing balloons. Authorities turned schools and various buildings in the empire into assembly sites. In the capital of Manchukuo, students of Hsinking Shikishima Higher Girls' School were among the young Japanese women mobilised to make balloons.[10]

Part of the process was inflating the spherical balloons – 10 metres in diameter – to check for leaks. This required large buildings. In Tokyo, the Kempeitai appropriated for the balloons the Nippon Theatre, Kokugikan Sumo Arena, Kokusai Theatre, and other sites. In those buildings, according to Major General Sato, even some of the capital's 'beauties of the stage' joined school girls in assembling the balloons. Even after taking over various buildings, the Japanese Army needed still more. Many new structures were constructed. The Army placed large orders for traditional Japanese paper and for konjak. Civilian manufacturers of such everyday products as Japanese writing paper and the translucent paper panels that covered the lattice frames of traditional Japanese doors and windows curtailed or halted production. Noodles and other products made of konjak disappeared from stores and dinner tables.[11]

Army officers managed the young women with patriotic appeals and dire threats. Tanaka Tetsuko was among the students drafted into service in August 1944 from Yamaguchi Higher Girls' School in southwest Honshu. She first worked at a factory built on the school's grounds. A military officer from Kokura Army Arsenal at the start sought to instil 'a sense of mission' in the girls by informing them that they would be assembling 'secret weapons' to strike the United States. She and her classmates – 150 girls – then went in January 1945 to the arsenal to work. An officer met them for their first shift, told them that the weapons they were making would defeat the United States, and led them in inciting the Imperial Rescript to Soldiers and Sailors to instil in them a sense of loyalty and devotion. Officers overseeing production at various sites would also warn those making the weapons against telling anyone what they were doing. One Kempeitai officer at Tokyo's Kokusai Theatre threatened execution for anyone, along with the offender's family, caught revealing what they were doing.

Teams of as many as 200 young women, each team working a twelve-hour shift, could produce perhaps ten balloons a day in a painstaking process of production by hand. The women, their long hair drawn back, wore headbands emblazoned with the Japanese rising sun or some patriotic slogan. Hairpins were forbidden and fingernails were cut short to prevent balloon punctures. Tanaka Tetsuko recalled in Kokura the 'deafening' factory where she worked, with plies of paper glued together and resting on metal frames that revolved around drying machines. While some workers pasted, other young women stitched together by hand the paper panels. Tanaka remembered that 600 sheets of paper, each approximately 1x2 metres, were required to make a single balloon.

Strenuous work, insufficient food, and spartan living quarters sickened many of the young women. Malnutrition, nervous breakdowns, and pulmonary tuberculosis were widespread. One young woman working in Nagoya recalled that she fell ill from beriberi and stopped menstruating from the stress of twelve-hour work shifts, a poor diet, and lack of sleep due to enemy air raids. Bitter cold in the dormitories and not enough hot water for the baths were part of the harsh environment. Some programme supervisors, unable to improve working conditions or provide nourishing food, gave the young women stimulants to help them work through their shifts. Tanaka recalled hearing in regard to the uppers given the girls in Kokura that Kyushu Imperial University was developing the pills to keep pilots awake and that the young women at Kokura Army Arsenal were 'testing' them.[12]

Launching Attacks

Beyond developing the bombing balloons, Noborito's First Section also had a hand in launching them. Kusaba and Takeda scouted candidate launch sites in northeast Honshu's Tohoku region and elsewhere on the main island. They excluded from their itinerary Japan's northernmost main island, Hokkaido, even though launch sites on the island's east coast offered the shortest routes to North America. Military leaders dreaded Moscow declaring war on Tokyo in the event of bombing balloons going off course and striking nearby Soviet territory. Kusaba settled on three sites. In the Tohoku region, he chose Nakoso, Fukushima Prefecture. In the Kanto region, closer to Tokyo, he picked out Ichinomiya in Chiba Prefecture and Otsu in Ibaraki Prefecture. The Army then built hydrogen storage structures, installed equipment for inflating the balloons, and made other preparations at the launch sites.

The time to launch drew near. In September 1944 came an order to organize a new balloon regiment under the Army General Staff. By late October, the military had prepared high-explosive bombs and incendiaries. Teams of young women had stitched together balloons for the first launches of the bombing campaign. On 25 October, General Umezu Yoshijiro, chief of the Army General Staff, issued the order to Colonel Inoue Shigeru, the new balloon regiment's commanding officer, to bomb the United States from November to the spring of 1945. Colonel Inoue's regiment was to launch approximately 15,000 balloons that all together were to carry some 7,500 15kg explosives, 30,000 5kg incendiaries, and 7,500 12kg incendiaries. Classified a secret within the military and hidden from the public, the launches would be called 'tests' and, weather permitting, were to take place at dawn, dusk, and night.[13]

On 3 November, the birth anniversary of the Emperor Meiji, the IJA launched from Ichinomiya its first bombing balloons. All three sites were operating from 7 November. General Sato, eager to strike back at the United States for bombing Japan, visited the launch sites, offering whiskey and words of encouragement to the officers and men. As the bombing balloons flew one day after the next towards North America, the Army sought to discover what damage its campaign was causing. Sato later wrote that the IJA monitored foreign radio broadcasts and exploited 'other' means, possibly a reference to communications intelligence, Japanese agents or both, to find out what what was happening in the United States. When a report reached him around the beginning of December of a disturbance somewhere in California resulting from a fire of unknown origin breaking out among trees in a windbreak, Sato 'leapt for joy', convinced that the news pointed to a successful bombing.[14]

Fireballs in the Sky

Fireballs appeared in the skies of the western United States. A woman on a ranch in Colorado witnessed within days of the first launches a sudden burst of flame in the sky and described it as 'a big brilliant ball of fire, about the size of the moon.' Other people in other places saw similar flashes of fire. Mysterious bits of Japanese paper were found in the streets of Los Angeles. The incidents seemed unreal. The Colorado woman who informed a government official of what she had seen told him that friends whom she had told thought that she had been 'having hallucinations'.[15]

Officials began recovering material that pointed not to hallucinations but to real weapons. On 4 November, a balloon was fished from the water off San Pedro, California. In the middle of the month, the US Coast Guard recovered a 'fragment of the envelope and some of the attached gear' of a balloon from the ocean near Hawaii. The next day, another balloon was found far inside mainland America, near Kalispell, Montana.

Navy personnel shipped the first recovered materials to the Air Technical Intelligence Command in Anacostia, District of Columbia, which determined that it was a Japanese meteorological balloon. Authorities also sent under armed guard to Anacostia the gathered fragments of the Montana balloon. On 6 December, nearly three years to the day of the Japanese attack on Pearl Harbor, Wyoming residents heard 'two heavy explosions' near the town of Thermopolis. One witness spoke of seeing a parachute or balloon falling to earth after hearing the explosions. A bomb disposal officer of the Army's Seventh Service Command examined the discovered bomb fragments and, according to a US military intelligence report, 'tentatively identified these parts as a 15kg Japanese fragmentation bomb of a type which would be used in aerial bombardment for anti-personnel purposes.'[16] The US mainland was under attack.

To defend the country, US officials needed to understand the bombing balloons. Did they come from land, by submarine or from ships? How did they operate? What were their capabilities? What did Japan hope to accomplish in launching them?

Officials began to solve the puzzle by gathering more fragments and, in January 1945, a complete balloon captured in flight near Crater Lake, Oregon. Chemical analysis of the balloon's hydrogen pointed to inflation at a 'large, land-based factory,' according to Lincoln LaPaz, wartime technical director of the Operations Analysis Section, Army Second Air Force. Dr LaPaz also noted that an examination of the ballast sandbags showed that the sand came from Japanese beaches. More evidence came from American B-29 pilots who returned from bombing runs against other targets in Japan with news of 'a novel "antiaircraft" device consisting of huge, shining spheres seemingly ascending to meet the American planes.' The devices that they had seen, deemed ineffective against the bombers, were bombing balloons on their way to the United States.

As they collected more and more pieces of the bombing balloons, officials came to understand the genius of their construction and the difficulty of defending the country against them. LaPaz wrote that he and his colleagues 'could not help admiring the ingenuity that had gone into the making of the balloon and its intricate cargo.' The technology was inexpensive, yet impressive. 'The balloon envelope was made of five layers of long-fibre rice paper glued together with hydrocellulose – as inexpensive a combination as could be imagined. Yet those thin paper envelopes leaked less than one tenth as much as our own costly rubberised-fabric balloons!' A report from the Naval Research Laboratory, dated 15 January 1945, described the balloon found off Hawaii as carrying 'an ingenious ballast release mechanism' that would enable long-distance bombings from Japan and a self-destruct mechanism that served 'apparently to avoid recovery of the equipment by the Allies or to avoid detection of the operation.' Defence against the paper weapons would be difficult. Dr LaPaz wrote that the 'crude radar' in service in 1945 would perform poorly against a weapon with so little metal in it.[17]

Japanese capabilities and intentions concerned the US Army. A memorandum of 26 December for the chief of the Military Intelligence Service (MIS) had referred to 'a bomb tentatively identified as Japanese.' The recommendation of the document, signed by two top intelligence officers in the War Department, Major General Clayton Bissell and Brigadier General John Weckerling, was that the MIS gather 'all available information on Japanese balloon activities and equipment' and determine the new weapon's capabilities. The report indicated five specific capabilities of concern: '1) Espionage or sabotage agents; 2) Lethal devices; 3) Bacteriological warfare materials; 4) Forest fire incendiaries; 5) Antiaircraft devices.' The memorandum also called on the MIS chief to 'study or evolve suitable countermeasures' and coordinate with 'interested agencies, including Army Air Forces, Navy Department, Western Defense Command, the Chief Signal Officer, New Developments Division, WDGS, Air Technical Intelligence Command and the Office of Scientific Research and Development.'[18]

Various government organisations would pass intelligence on the new weapons to the Western Defense Command (WDC), responsible for defending the western United States against the incoming intercontinental weapons. Military officials established within MIS a team to investigate the bombing balloons in a project assigned the number 1337-A. Colonel Eric Svensson, a Japan specialist, chaired the team, which included a lieutenant colonel serving as a 'domestic specialist' and comprised several branches: Topographic, Domestic, Military, Sociological, Scientific, and Economics. Scientific Branch's duties included preparing interrogation materials for Japanese prisoners of war and looking into 'bacteriological warfare possibilities; hazards to humans, animals, and plant life.'[19]

Propaganda from Japan, Censorship in North America

Begun as a secret programme, Fu-go became grist for the Japanese propaganda mill. On 17 February 1945, Tokyo Radio reported learning 'only yesterday' via Chinese media of the bombing near Kalispell, Montana, on 6 December 1945. The announcer further reported that the Japanese government had announced attacks 'by means of a new secret "V" weapon, Japanese Fireflies, against our enemies on the North American continent. No longer do the American people enjoy freedom from air attack.' Mixing fact and fiction, the announcer claimed 10,000 deaths in North America. In fact, there had been no deaths at all. The broadcast also claimed that the balloons could 'carry several persons, and the day is not far distant when we will land several million Japanese troops on American soil.' Kusaba recalled that he and others at Noborito and elsewhere were 'overjoyed' on hearing that the bombing balloons had reached North America.[20]

The US government then moved to prevent Japan from gaining more information on the results of their bombings. The Office of Censorship (OC), established within days of the attack on Pearl Harbor, oversaw the voluntary Code of Wartime Practices for the press to follow. To deny the Japanese information that their intercontinental weapons were arriving in great numbers, the OC asked the press and radio to report only incidents for which the War Department gave official information. Editors and broadcasters complied. Balloons continued after the Kalispell incident to reach the United States and drop their payloads of explosives and incendiaries, but US media compliance with the OC Code deprived the Japanese Army of information that they sought each day in monitoring the airwaves and poring over the foreign press. Canada, too, contributed to keeping Japan in the dark. While Canadian military authorities did pass by word of mouth limited information to residents in areas where the balloons were sighted, Canadian mass media kept quiet. The blackout frustrated Japanese officers. Kusaba recalled that after the elation over the early press report, 'We frantically searched the media of various countries, the United States, the Soviet Union, and China, but were ultimately unable after that to find any reaction until the end of the war.'[21]

Gathering Intelligence, Preparing Defences

Responsibility for gathering intelligence on the bombing balloons and countering them fell to the Army's WDC, with its headquarters at the Presidio, San Francisco, and the Navy's Western Sea Frontier (WSF), headquartered in the same city. These two entities in turn coordinated their activities with other military, naval, and civilian agencies.

The area commands worked to draft and execute plans to meet actual and potential threats posed by the bombing balloons. The Japanese weapons floated freely, without guidance mechanisms. Their anti-personnel explosives thus posed less of a danger over the wide expanses of the West than did the incendiaries, which could start forest fires, a particular threat in the dry season. The WDC, working with the Army's Ninth Service Command and Fourth Air Force, participated in the Firefly Project, in which thousands of troops were deployed and aircraft readied for fighting forest fires.

The greatest potential threat was that of balloons carrying biological weapons, which could prove devastating even without accurate bombing. The US response was the Lightning Project. Officials involved instructed agricultural and health officials to monitor crops and livestock for unusual outbreaks of disease. They also deployed decontamination materials in the western United States in case of detected incidents. Outside WDC and WSF, the Chemical Warfare Service (CWS) had 'responsibility for the study of possible carrying of biological agents by these Japanese balloons… with the collaboration of The Surgeon General's Office in matters of defence and clinical medicine', according to a secret memorandum of 12 February 1945 from the Adjutant General's Office in the War Department. As such, the CWS took possession in the nation's capital of specimens collected from recovered bombing balloons for examination for biological agents.[22]

American authorities were right in their concern regarding the BW threat. Kusaba wrote late in the 1970s on IJA thinking behind launching balloons with biological payloads against the United States:

> Everyone takes for granted that what are most effective are those things whose force remains after they reach the target, such as propaganda leaflets, as well as bacteria and harmful insects directed against animals and plants, and in particular those things whose force expands. In particular, when bacteria and pests go to virgin land, they face no enemies, which results in unexpected propagation. Research on such things was conducted with this in mind. When the plan of attack was established, however, there came an order from above not to use them. Therefore, the decision was made to drop mainly incendiary bombs.[23]

Nuke, Paper, Lead

On 10 March 1945, around half past three o'clock in the afternoon, a farmer witnessed a balloon coming to earth on his farm near Toppenish, Washington. The balloon, the farmer reported to an FBI special agent later that day, struck a high-tension power line of the Bonneville Power Administration before bursting into flames. Three attached incendiary bombs hit the ground but stayed intact. In a memorandum later that month to Major General Leslie Groves, director of the Manhattan Project to develop nuclear weapons, Colonel F.T. Matthias, area engineer in the Corps of Engineers, explained that the balloon's contact with the wires had 'apparently caused a line surge.' The incident had created an outage 'sufficient to trip the vertical safety rods in the piles.' Plant personnel reportedly had B pile working again in 10 minutes, and D pile was up in 12 minutes, but it took them over an hour 'to get the F pile back under power due to difficulty in one vertical rod.' A paper balloon had brought to a temporary stop the nuclear reactors of the Hanford Engineer Works, which produced plutonium for the Manhattan Project.[24]

As paper beats rock in the popular children's game of rock, paper, scissors, so paper that afternoon – if only briefly – beat nuke. Nearly five months later, however, a plutonium bomb detonated over Nagasaki destroyed much of that Japanese city. Some 40,000 people died in the attack; tens of thousands more died later from injuries and radiation.[25]

Later that March afternoon, a Hanford security officer and an Army MP spotted an inflated balloon, driven by the wind, moving with its undercarriage towards the same transmission line that had been struck near Toppenish. The two men drew their service arms and shot the balloon to deflate it and stop it from coming into contact with the line.[26] By firing metal through paper, the two men prevented a possible repeat of paper beats nuke.

In the course of the war, 'several' bombing balloons 'came down near the Hanford atomic bomb factory.'[27] Fortunately for plant personnel, there were no direct hits.

CHAPTER 5

COUNTERFEITING CURRENCIES, FORGING DOCUMENTS

> Economic warfare, especially the subversive operation of counterfeiting and distributing an enemy country's currency on a large scale, thereby throwing the economy into disorder, is a concept in which one anticipates real and total warfare against a powerful enemy. As such, it was something that I could not even imagine at the initial stage of the clash between Japan and China. It was something impossible.
>
> – Colonel Yamamoto Kenzo, director of Third Section, Noborito Research Institute[1]

Wrecking the Enemy's Economy

Wars were once limited affairs, won or lost on fields of battle. With the rise of nations and the industrial revolution, destroying the other side's economy became part of war. Burning crops, destroying factories, tearing up rail lines, and blockading ports were overt ways of waging economic warfare. Counterfeiting enemy currency was a covert course of action.

Tokyo was not alone in turning to counterfeiting in the Second World War. Allied and Axis nations alike printed enemy currencies in secret. Engravers in the employ of America's OSS had completed work on counterfeiting the German Reichsmark notes and were nearing the end of their project to reproduce the Japanese yen. Tokyo surrendered before the bogus yen bank notes started circulating in Japan. The Germans, too, set up a clandestine operation. Using Jewish prisoners at a concentration camp in Sachsenhausen, the German officer in charge turned out British pounds and was overseeing work on the US twenty dollar bill when Berlin surrendered.[2]

Starting Down the Path

Japanese leaders by the summer of 1938 were eager to find a way to end the fighting in China. Following the clash the previous summer between Japanese and Chinese forces near the Marco Polo Bridge, Japanese had overcome Chinese defences in a number of cities but failed to compel the Nationalists under Chiang Kai-shek to surrender. When IJA forces took the Chinese capital of Nanking in December 1937, the Nationalists had retreated into the vast Chinese hinterland to continue the fight. Eager to end Nationalist resistance, the Japanese Army had considered poisoning Chinese leader Chiang Kai-shek.

Instead, Tokyo sought a solution in a clandestine operation to wreck the Chinese economy. In July 1938, a meeting of Japan's Five Minister Council – the Prime Minister and the Army, Navy, Finance, and Foreign Ministers – decided on several measures to end the fighting in China. Among them was the decision to study the feasibility of destroying the Nationalist currency: the fapi, also known as the yuan. China's legal tender since its introduction in 1935, the fapi had replaced various bank notes previously in circulation. While even private banks had issued their own notes prior to 1935, the Nationalist Government restricted that function to four banks: the Central Bank of China, the Bank of China, the Bank of Communications, and the Farmers Bank of China.[3]

Japan carried out both overt and covert attacks on the Chinese currency. As the Japanese Army took over major cities and rail lines in eastern China, Japan sponsored local governments, banned the use of the fapi, and favoured competing currencies or military scrip. Counterfeiting the fapi would be the covert way to undermine the currency. The goal was to introduce fake bank notes on a large scale to provoke ruinous inflation that would sap Chinese will and destroy the economic basis for resistance. In addition, the Japanese Army, would use counterfeit Nationalist currency to purchase the large quantities of local supplies need to sustain military operations. Colonel Yamamoto Kenzo, who would direct the covert counterfeiting operation, included in his postwar memoir a quote from the Chinese strategist Sun Tzu in regard to soldiers waging war in enemy territory: 'they rely for provision on the enemy.' The quoted except is part of a longer passage from Sun:

> They carry equipment from the homeland; they rely for provision on the enemy. Thus the army is plentifully provided with food.[4]

As Yamamoto wrote, Sun's concept of living off the land was 'the basic thinking of Japan's logistics strategy.' A Chinese academic described the strategy of undermining China's resistance while procuring local supplies with counterfeit bank notes as 'an extremely evil stratagem of killing two birds with one stone.'[5]

Yamamoto Arrives at Second Bureau

Yamamoto Kenzo, born in Tokyo in 1901, began his military career in 1921 in graduating from the Army Intendance School. Ordered to the 9th Infantry Division, he served in the Japanese military campaign in support of White Russian forces against the Communist Red Army in the Soviet Far East. The Japanese Army sent Yamamoto to the Tokyo School of Foreign Languages in 1929 for Russian and the Army Intendance School in 1931 for advanced studies before assigning him in 1933 to the Kwantung Army in Harbin. He spent two years in Manchukuo, compiling topographical information for military maps and studying Chinese and Manchurian currencies.

In March 1938, after a few months spent in Shanghai when the Japanese Army had mobilised troops as fighting spread in China after the Marco Polo Bridge Incident, Yamamoto started in Tokyo a new assignment in the military topography sub-section of AGS Second Bureau's 7th Section (China). There he learnt of earlier, failed attempts at counterfeiting Chinese currency. Having previously studied bank notes in Manchuria, convinced that counterfeiting could work, Yamamoto swayed his subsection's chief, Lieutenant Colonel Sakata Shigeki, who sent him that year to China to conduct a survey. Yamamoto looked into the use of fapi and other bank notes in traditional Chinese banks in the great commercial hubs of Canton and Shanghai, visited moneylenders in the British colony of Hong Kong, and entered gambling dens in the Portuguese colony of Macao to see how money was circulating and in what forms. He also learnt which Chinese, American, and British companies were printing the Chinese bank notes.

After returning to Tokyo, Yamamoto consulted Japanese paper and printing companies on technical issues in producing bank notes. He then drafted a plan of action and submitted it in the spring of 1939 to Colonel Watari Sakon, chief of 7th Section. Yamamoto's proposal then hit the desk of Colonel Usui Shigeki, chief of 8th Section, established the previous year to oversee the IJA's clandestine operations in China and elsewhere. After reviewing the plan, Usui directed Yamamoto to take it to a remarkable officer who had departed 8th Section to head the Military Affairs Section in the War Ministry's Military Affairs Bureau: Colonel Iwakuro Hideo.[6]

Iwakuro, as noted earlier, had played a driving role in improving Japanese military intelligence. He also acted overseas. In 1941, he joined Admiral Nomura Kichisaburo, Japan's ambassador to the United States, and Ikawa Tadao, a former official of the Finance Ministry, in Washington in negotiations with the United States to find a compromise that would protect Japanese interests in Asia. Ikawa described Iwakuro to American interlocutors prior to the colonel's arrival as one of 'the driving forces of the Army.' Ambassador Joseph Grew sent from his embassy in Tokyo to the Secretary of State in February 1941 a telegram referring to the army officer: 'Colonel Iwakuro, according to a reliable source, is one of the most important leaders of the young officers' group and has the complete confidence of the Minister of War.' When Tokyo chose war rather than submit to Washington's ultimatum to withdraw from the continent, Iwakuro commanded the 5th regiment of the Imperial Guards Division in the Malaya Campaign that ended in the Japanese capturing the British Far East bastion of Singapore. Wounded in the fighting, Iwakuro recovered to assume in April 1942 command of the IJA operation to undermine British rule in India.[7]

Yamamoto also saw something extraordinary in Iwakuro. At their first meeting, Yamamoto found him to be 'far from' the average military man, someone 'both keen of mind and flexible in character.' Iwakuro, convinced on hearing Yamamoto's plan and his declaration that he would 'give his life' for the operation, put him at the centre of it. A week later, Iwakuro called Yamamoto back to his office and ordered him to proceed to Noborito, where he would oversee the production of counterfeit Chinese bank notes. Iwakuro informed Yamamoto that he would be working on the project with Lieutenant Colonel Okada Yoshimasa, an officer in AGS Second Bureau's 8th Section. An operative by the name of Sakata Shigemori would run the project in China from Shanghai.[8] Yamamoto was transferred from 7th Section and given concurrent assignments to 8th Section and Noborito. His mission at Noborito was to produce counterfeit bank notes in Operation Sugi, the IJA project to wreck China's economy.

Lieutenant Colonel Okada, an infantry officer from IMA Class 36 (1924), was a military China hand who, fresh from an assignment in Japan's Asia Development Board, arrived in December 1938 for duty at 8th Section of AGS Second Bureau. He had been there only a few months when Iwakuro called him to his office to inform him that he was to work out a plan for a covert economic operation to end Chinese resistance. Iwakuro told Okada that Yamamoto was working to counterfeit Chinese currency and that there would be no more 'childish stuff' like the fake notes that Chinese had previously detected. Iwakuro was adamant, telling Okada: 'We're going to make real ones!'

Iwakuro's pick to run day-to-day operations from Shanghai, Sakata Shigemori, had worked over many years with Okada and other Japanese military officers as a China *ronin*. The term *ronin*, which in Japan's feudal era referred to a masterless samurai, meant in modern times the various Japanese in China and elsewhere overseas who, while holding no government position, carried out various missions for Japanese officials. Sakata had gone to China as a young man for adventure. Graduating from a university in Peking, fluent in Chinese, he had gone to work for the South Manchuria Railway Company, developed close ties to China's notorious Green Gang, and worked in the shadows for the Japanese Army. Transporting opium in IJA operations was one of his services. Lieutenant Colonel Okada had first encountered Sakata in Peking in 1927 while serving in China in the Japanese Army's Shantung Expedition. Okada viewed Sakata as a 'magnanimous, truly fine man.' That was their background when Okada, by then working in 8th Section, had called Sakata late in 1938 to the Imperial Hotel in Tokyo for a week of discussions on how to execute the counterfeiting plan.[9]

Yamamoto Arrives at Noborito

Prior to Yamamoto's arrival, Shinoda had been in charge of counterfeiting at Noborito. This particular area of covert activity, which started in Shinoda's laboratory at the ASRI, transferred to Noborito in 1939.

Shinoda was an expert in paper technology, but not in the printing of bank notes. He went about enlisting experts and recruiting new talent. For technical guidance, Shinoda went to the ultimate authority: the Japanese government's Printing Bureau, the producer of Japan's own yen bank notes. In April 1937, the Printing Bureau's Kawahara Hiroma joined Shinoda's laboratory to work on inks.[10]

The Tokyo School of Industrial Arts, then the only technical school in Japan with a printing section, was a source of new talent. Oshima Yasuhiro, a 1939 graduate of the school, applied to work at the ASRI. Two motives moved him to do so. He wished both to learn new technology and to stay far from the fighting in China. After the Kempeitai had gone over his background, Oshima wound up working in Noborito's Third Section as an engraver.

Studying the inks, paper, and techniques used in manufacturing Chinese bank notes was the start of the counterfeiting effort. Yamamoto, with Iwakuro's approval, then enlisted the aid of the Tomoegawa Paper Company soon after his arrival at Noborito. At the end of 1939, Yamamoto took some counterfeit bank notes with him to Shanghai for field testing. By that time, Lieutenant Colonel Okada was operating as senior staff officer in the intelligence section of the China Expeditionary Army Headquarters in Nanking, directing an operation in Hong Kong before Japan's invasion of the British colony, and directing a local SSA in Shanghai: the Matsu Agency.[11]

Sakata was directing his own organisation, the Sakata Agency, to carry out IJA operations. In a test of Japanese counterfeit bank notes, Iwakuro recalled, Sakata used a mix of real and counterfeit notes to buy several dozen handkerchiefs and other items at Shanghai's Sincere Department Store. He later presented Iwakuro with the handkerchiefs as souvenirs of his first successful purchase. Sakata was also able to purchase some platinum. In other cases, however, Chinese bankers in Shanghai rejected the bank notes as counterfeit. Noborito's Third Section would have to do better.[12]

Assembling equipment and personnel, experimenting with different techniques, Yamamoto and his men continued working towards Iwakuro's goal of printing 'real ones.' In 1940, Noborito acquired at great cost from a German firm sophisticated four-colour equipment for collect-printing, also known as Sammel printing. As war had erupted in Europe by then and the port of Hamburg was unavailable, IJA routed the equipment south to Naples and shipped it from there to Japan.

After Japan had declared war against Great Britain and the United States on 8 December 1941 and taken the British colony of Hong Kong on Christmas Day, men of Noborito's Third Section flew to the colony to collect some special spoils of war. Okada, already in Hong Kong as part of an advance team, reported to Yamamoto that, with Hong Kong's paper and printing facilities in Japanese hands, his team had found real Chinese bank notes as well as the materials and equipment to make them. Yamamoto immediately flew to Hong Kong to see the captured items, as did Major Kawahara Hiroma, a group director in Third Section, and his subordinate Oshima. After taking possession of actual bank notes, printing equipment, and plates from Nationalist printing plants in Hong Kong – those of the Chung Hwa Book Company, the Ta Tung Book Company, and the Commercial Press – and sending them to Noborito, Yamamoto and his men were no longer producing unsatisfactory counterfeits. As Okada put it, the counterfeit bank notes in effect became indistinguishable from legitimate ones.[13] Noborito in the first half of 1942 had reached Iwakuro's ambitious goal of producing 'real ones.'

From Inside the Barbed Wire Out Into China

Yamamoto's counterfeiting unit was a secret even inside Noborito. Within the perimeter of the guarded facility were barbed-wire enclosures for buildings of Third Section. Kuba Noboru, director of Second Section's Group 7, recalled after the war that it was 'difficult' even within the Noborito compound for personnel 'to know what the various laboratories were researching.' Nor, he noted, did he see any evidence 'that people were trying to find out.' Kuba once entered a fenced-off building between Group 7 and Second Section's main building in search of a men's room, only to encounter a Kempeitai sergeant demanding to know his name and what he was doing there. Kuba later received from Noborito's director, Shinoda Ryo, a gentle warning against visiting that building again. Kuba sensed that 'behind the director's smile, he was satisfied that a secret of the institute was being kept.'[14]

Noborito's Third Section comprised three groups. North, under Major Ito Kakutaro, was responsible for paper production. Centre under Major Okada Masayuki, handled analysis, recognition, and printing ink. South, under Major Kawahara Hiroma, took care of engraving and printing.

The group names came from their location relative to one another within the grounds of Noborito.

Yamamoto had official, commercial, and student cooperation. The government's Printing Bureau gave technical guidance and sent technicians to Third Section. The Toppan Printing Company worked with Yamamoto's men to make the plates. Tomoegawa Paper and other companies assisted in producing the paper. Yamamoto also brought 25 female students from a local high school into the operation. Their task was to give the crisp new notes a well-used appearance similar to that of legitimate bank notes already in circulation. The girls gave the new notes a weathered look by turning them in a concrete mixer with dirt, garlic, pig fat, and other materials.[15]

Yamamoto enjoyed the highest backing. The War Ministry in March 1941 had ordered Noborito to provide Army Minister Tojo Hideki with a briefing on the counterfeiting project. Yamamoto and an officer from the War Ministry's Military Affairs Section briefed Lieutenant General Tojo and Vice Minister Lieutenant General Anami Korechika, showing them examples of actual bank notes and Noborito's counterfeits. Tojo, much impressed, subsequently instructed Military Affairs Section to make sure that Yamamoto had the budget he required. Third Section was so much more flush with funding than other parts of Noborito that Shinoda at times asked Yamamoto to transfer some of his money to other sections. Tojo also received reports, highly restricted, on the operation's progress via the Military Affairs Section.[16]

After production, the next step in Operation Sugi was transporting the bank notes to Shanghai. Non-commissioned officers (NCOs) who had graduated from the Nakano School handled that task. The IJA sent nine Nakano NCOs to Noborito. All but one worked for Third Section. Three of the eight – Kukita Yukiho, Shiraishi Takao, and Kumamoto Takeharu – shuttled between Noborito and Shanghai as part of Operation Sugi. A fourth operative, Kuritaguchi Shigeo, later became an aide to Yamamoto. All the Nakano NCOs at Noborito were formally assigned to 8th Section of AGS Second Bureau. Kuritaguchi alone had a concurrent assignment to Noborito.

Kukita described after the war the transport of counterfeit bank notes to China. At Noborito, the notes were bundled by denomination – 10, 20, 50, and 100 yuan – and packed into wooden boxes, each containing 200,000 yuan of counterfeit notes. On each trip, two or three of the Nakano men, dressed in plain clothes to avoid drawing attention, would take five crates – one million yuan – by passenger train to Nagasaki. They would then board the SS *Shanghai Maru*, a ship of the Nippon Yusen Kaisha (NYK) Line that sailed between Nagasaki and Shanghai.[17] On arriving in Shanghai, the men would disembark without interference from the Kempeitai. The team would then enter the city's former French Concession, proceed to Yuyuan Road, and deliver the crates to Sakata's organisation at the Tian Mansion, where they would stay while in the city. On the grounds of the mansion stood a storehouse for keeping the counterfeit bank notes. Wives of local Japanese residents would unpack the crates there and bundle the Noborito's counterfeits with real bank notes for future use.[18]

The Japanese Army was running its counterfeiting operation in Shanghai from a spacious mansion thanks to Sakata Shigemori's contacts with the Chinese criminal underworld. The property belonged to Tu Yueh-sheng, head of the fearsome Green Gang, which worked with both the Japanese Army and the Chinese Nationalists. With roots reaching back in history to underground movements against China's Manchu rulers in the late Ching Dynasty, the Green Gang from their base in Shanghai engaged in opium smuggling, gambling, and other disreputable activities. Sakata had cultivated ties to the organisation over the years. As a demonstration of his bona fides, he had even married the daughter of a major figure in the gang.[19]

The Green Gang served the Japanese Army both as a bridge to the Nationalist leadership in Chungking and as an organisation to move Noborito's counterfeit bank notes from the storehouse on Tu's property into the Chinese economy. Tu himself had withdrawn to the nationalist capital of Chungking, leaving trusted lieutenants to oversee his Shanghai affairs. The Hua Shin Company, located on another of the Sakata Agency's properties, was particularly involved in putting the fake money into circulation. Previously known as the Cheng Ta Company, a joint venture between Hsu Tsai-cheng, Tu's representative in Shanghai, and Sakata, the trading company also connected the Japanese Army to Chungking, serving as a channel in IJA efforts to end the conflict through 'peace operations'. Hsu was also president of the Min Hua Company, previously known as the Ta Chi Company, a trading company that operated in areas outside Japanese control. The Green Gang's ventures would use counterfeit bank notes to purchase in 'enemy territory' such commodities as salt and lumber, which would go via the Matsu Agency into IJA depots.[20]

The Japanese Army also sent Noborito's bank notes from the Sakata Agency to other IJA organs. Kukita and other Nakano NCOs from Shanghai transported the notes to the Matsu Agency's Canton branch, Shorindo, which operated under Itagaki Kiyoshi, a talented China *ronin* who had joined the Matsu Agency at the age of nineteen.[21] Shorindo used the counterfeit notes to procure gold, tungsten, petrol, and petroleum. Itagaki recalled receiving bank notes worth an 'astounding' 80 million yen. He was also impressed by the quality of Noborito's counterfeits and noted with admiration that the merchants in Macao, an international market for metals, never seemed to have detected the fake notes.

The Ume Agency under Major General Kagesa Sadaaki, as the IJA military advisor to the government of Wang Ching-wei in Nanking, also used Noborito's bank notes to bankroll his organisation's operations.[22] Even the IJN benefited from the IJA's counterfeiting operation. Kodama Yoshio, a China *ronin*, ran the Manwa Trading Company, which procured materials for the Japanese Navy. He also established his own Kodama Agency. Short of funds, Kodama turned to Okada, who supplied him with Noborito bank notes for his purchases.

Finally, the Japanese Army used Noborito's bank notes to procure supplies for its troops when advancing deep into enemy territory in 1944 in the massive Operation Ichi-Go to develop rail lines and roads under Japanese control and to capture Chinese airfields from which American bombers were attacking Japanese targets.[23] Japanese commanders in the campaign acted in line with the ancient strategist Sun Tzu's formula of relying for their supplies on the enemy. The twist was that the Japanese Army in this case was not seizing supplies but paying for them with counterfeit bank notes utterly genuine in appearance.

Counterfeit Rupees for Operations in India, Trial Production of US Dollars

Beyond China, Noborito's Third Section produced counterfeit rupees for IJA operations against British rule in India. The Japanese Army had gained early in the war a pool of Indians for such operations. As Japanese troops made their way down the Malay Peninsula towards their ultimate goal, the British bastion at Singapore, Major Fujiwara Iwaichi and his F Agency succeeded in persuading surrendered Indian soldiers to induce further Indian surrenders in British ranks. After the conquest of Singapore in February 1942, with some 50,000 Indian soldiers among the surrendered forces, and the fall in March of Rangoon, the capital of British Burma, the Japanese Army had many resources for propaganda and clandestine operations against the British in India.[24]

Late in March that year, Iwakuro took the baton from Fujiwara to launch from Saigon his Iwakuro Agency to undermine British rule in India. He soon advanced the headquarters of his large organisation to the Thai capital of Bangkok. Iwakuro's agency broadcast propaganda, much of its news featuring the Indian National Army (INA) and directed at British India; gathered intelligence; and conducted clandestine operations in India.[25]

Iwakuro called on Noborito to make Indian rupees for his operations. Noborito Third Section's Colonel Yamamoto found the task 'not that difficult.' Kukita, one of the Nakano NCOs who shuttled between Japan and Shanghai with counterfeit Chinese bank notes, told a postwar historian that he also made multiple trips with Noborito rupees to the Iwakuro Agency in Bangkok.

The rupees that Third Section produced were a form of 'logistical support' for the operations, according to Kukita, who was one of several Nakano NCOs who secretly brought over 50 crates of counterfeit Indian bank notes via Saigon and Phnom Penh into Bangkok for operations. Iwakuro sought to undermine British rule by inserting Indian agents by air and sea into India. He explained that he made 'considerable' use of counterfeit rupees to gather intelligence in those agent operations. According to a US intelligence study of 4 September 1945, the Hikari Agency sent 'large numbers of native agents' into India on short-term missions that netted 'a limited amount of accurate information.'[26]

Noborito also played a role in Indian propaganda. With the 1943 arrival in Bangkok of Subhas Chandra Bose, a prominent independence activist who had been broadcasting earlier in the war by radio from Berlin against British rule in India, the Hikari Agency sought to enhance its operation by elevating his status, proclaiming him INA commander and Indian leader in exile of the Provisional Government of Free India. In line with this propaganda campaign, the War Ministry tasked Noborito with producing a new currency for Bose's government. Third Section already had more than enough work in counterfeiting Chinese yuan, according to Yamamoto, but Noborito had to comply. Yamamoto arranged for Third Section to supply the paper and Toppan Printing to design and print the new currency. In the end, however, the Bose government's bank notes 'never saw the light of day.'[27]

Yamamoto's Third Section also made an effort to counterfeit US dollars in efforts, but their work reportedly yielded little. Iwakuro admitted years after the war that Noborito had tried making dollars but said that the counterfeit greenbacks 'did not turn out well.' A history of the Nakano School written by its veterans also suggests that counterfeiting the US dollar would have been 'extraordinarily difficult.' Even if Noborito had produced worthy counterfeit bank notes in large volume, according to the insider history, there was no viable way to introduce the bank notes in large numbers into the United States. Yamamoto denied that there had been any plan to drop counterfeit dollars by balloon on the United States. The effort reportedly ended with little more than some field testing of fake dollars in Macao.[28]

Authenticating Agents With Passports, Documents

Other than counterfeiting cash, the Noborito Research Institute produced passports and other documents to make the false identities of agents appear real as they crossed borders and operated in enemy territory. In this, Colonel Yamamoto's Third Section worked in an area common to all countries with developed intelligence services: agent authentication. Stanley Lovell, director of OSS Research and Development Branch, recalled that upon entering government service:

> I decided that the very first job to be done was the organisation of a plant for documentation – a fascinating, meticulous, deadly business indeed. It was obvious that any spies or saboteurs OSS placed behind enemy lines would have short shrift unless they had perfect passports, workers' identification papers, ration books, money, letters and the myriad little documents which served to confirm their assumed status. These are the little things upon which the very life of the agent depends.[29]

Yamamoto wrote that Third Section, 'it goes without saying', produced not only counterfeit money but produced 'on request' the passports 'necessary for intelligence and clandestine operations' in various countries.[30]

Passport work was perhaps the activity that most weighed on the minds of Third Section's members. Oshima Yasuhiro recalled the tension that they felt in making such documents. He explained: it was one thing if a counterfeit bank note were detected, but a passport that failed the test meant the death of a Japanese agent. Most such IJA agents were White Russians whom the Japanese Army would parachute into Siberia with fabricated documents and radios to communicate their findings. The Kwantung Army in Manchukuo called on Noborito to produce Soviet documents for their agent insertions. According to Yamamoto, the Japanese had been relying on Soviets passports provided by the Germans, but the watermarks and ink failed to withstand inspection. Every agent that went into the Soviet Far East with them was caught. Berlin then sent equipment for producing passports, from which point Noborito made its own.[31]

CHAPTER 6

WITHDRAWING FOR BATTLE, DESTROYING EVIDENCE

> Take measures immediately to destroy all evidence of special research that would be to our disadvantage if the enemy were to obtain evidence of it.
>
> – *Special Research Disposal Procedures*, Military Affairs Section, Military Affairs Bureau, War Ministry, 15 August 1945[1]

Japan was waging in early 1945 an increasingly desperate fight against enemy forces that were advancing ever closer to the home islands. The Japanese Army had succeeded in its massive Ichi-Go operation of April to December the previous year in destroying a number of airfields in China from which US bombers had been striking Japan, but Japan's loss of the Pacific island of Saipan in the summer of 1944 had given the United States a new base from which to launch devastating bombing attacks. In February 1945, two Marine divisions had gone ashore on the island of Iwo Jima to seize the strategic island as another base for air operations against Japan.

As Japanese military leaders prepared for an anticipated decisive battle in Japan's home islands, officers and technicians of the Noborito Research Institute were preparing new sites in the mountainous interior of Central Japan for developing new weapons and producing materials for the fighting planned for late 1945.

Operatives on Remote Islands

As part of its planned defence, the Japanese Army sent personnel with spy gear and special weapons from Nakano to execute operations on remote islands. Japanese commandos and soldiers on those islands were to organise local residents to fight the invaders. They were to report on enemy movements. If an island were overrun, the commandos would continue to lead operations until relieved of their duties.

In late 1944, the Japanese Army ordered 11 Nakano School officers to smaller islands south of Kyushu in the Ryukyu chain, which included the main island of Okinawa. Sergeant Kikuchi Yoshio, who went ashore at Iheya, an island northwest of Okinawa, brought with him special explosives packed in toothpaste tubes, weapons developed at Noborito, which he hid in a mountain hut. Sergeant Sakai Kiyoshi, who landed on Hateruma with his own allotment of Noborito equipment, had trained to inject bacteria from Noborito's special fountain pens to foul enemy water supplies.[2]

The Japanese military operatives who arrived at remote island outposts also brought with them radio equipment developed at Group 2 in Noborito's First Section under the direction of Major Takano Yasuaki. The Noborito radio equipment, in the words of a postwar US Army report, was 'specifically designed for use by assault troops and to this end was made as small and light as possible.'[3] Second Lieutenant Tokutomi Yukio, sent to the Amami Islands southwest of Kyushu, described his radio set, 'invented at the Noborito Research Institute,' as 'about the size of two connected cigarette boxes.' He explained that with the 'excellent' six-volt device, featuring a hand-cranked generator, operators could communicate at a range of 1,500 kilometres, allowing communication between operatives on remote islands and the higher chain of command in the home islands.

Tokutomi and his comrades could receive orders, report local conditions, and follow reports on the fighting elsewhere, including General Ushijima Mitsuru's last stand on Okinawa. Lieutenant Colonel Yoshinaga Yoshitaka, involved in the development of military equipment as a technical officer at the AOAH, recalled that Noborito's radio equipment included sub-miniature vacuum tubes, which were the state of the art prior to the invention of transistors, and that radio operators at Iwo Jima were able to keep contact with the home islands 'until the end.' The technician Kitazawa Ryuji, impressed with Takano's accomplishment, recalled that radio operators on Iwo Jima kept in radio contact for six months after the Japanese island fell to US forces.[4] Given the end of organised Japanese resistance on 25 March, assuming Kitazawa's memory to be correct, Japanese soldiers were using Takano's radios to communicate with officers in the home islands well after the Emperor announced by radio on 15 August Japan's decision to end the war and even after Japanese representatives signed the articles of surrender on 2 September 1945 aboard the USS *Missouri*.

Withdrawing Inland

Desperate circumstances in 1945 led to desperate measures to tap final resources in defence of the Japanese Empire. In Manchukuo and China, Japanese barely out of high school and men in middle age alike were drafted for military service. The Noborito Research Institute, too, took green youths into service. Shinta Shoji, sent to Kyoto Imperial University at the age of 17 for a half year of technical education, reported to Noborito in April. A sergeant processing Shinta's paperwork glanced at his date of birth and shouted to everyone in the room, 'Hey, we've got someone here born in the Showa era,' referring to the reign of Emperor Hirohito, also known as the Showa Emperor, which started in 1926. 'Now that we've got Showa-born people coming, it's all over for Japan!' The young Shinta, assigned to Fourth Section, observed defective material from the bombing balloons recycled for women at Noborito as 'balloon raincoats' and saw senior colleagues dismantling and boxing equipment for withdrawal to Nagano Prefecture, a mountainous prefecture of central Japan.[5] Fourth Section's Sugita Masami knew that equipment production at Noborito's equipment had by then declined 20 per cent from its peak in the years 1941-1943 due to material shortages. He could see that, due to the extensive military draft, only personnel were in abundant supply.[6]

As Japan came under increasingly destructive air raids, Japanese leaders made plans to withdraw from Tokyo and other coastal areas. Many of those withdrawal sites were in mountainous central Honshu, Japan's largest island. Late in 1944, digging began under IJA orders for an underground Imperial General Headquarters extending under three mountains of Nagano Prefecture. The site was to include living quarters for the Imperial Household and communication facilities for directing military forces and broadcasting propaganda to the people. Mitsubishi Heavy Industries, builder of the Japanese Navy's famed Zero fighter, and other key military corporations relocated factories to Nagano and nearby prefectures. They built some of their facilities under ground. The IJA Nakano School left Tokyo in March 1945 for Gunma Prefecture.[7]

In September 1944, Shinoda ordered key subordinates to prepare their sections for withdrawal. He and Yamada Sakura, director of Second Section, chose to relocate Noborito Second and Fourth Sections to the Ina Valley of Nagano Prefecture, nestled between Japan's Central and Southern Alps. They settled on the village of Nakazawa, suitable for its proximity to the Matsushiro Imperial Underground Headquarters and the Nakano School. Some members of Noborito's Third Section decamped for Takefu, Fukui Prefecture, the hometown of Third Section's Major Ito Kakutaro, who became chief of Noborito's new branch factory there. First Section and another part of Fourth Section withdrew to the village of Ogawa in Hyogo Prefecture. The withdrawal of much of the 9th ATRI from its hilltop site southwest of Tokyo took place over several months. There remained at war's end a 'Noborito Branch,' together with three new branches at their new sites.[8]

The Noborito Research Institute at the war's end had grown in dramatic fashion from its initial 18 members. An AOAH report dated 31 August 1945 put institute personnel at 861 persons, of which 131 were commissioned military officers and senior civil servants, 112 non-commissioned officers and junior civilian officials, and 616 second-class civilians and workers.[9]

Relocating Noborito units as the bombs were falling and American forces were approaching the home islands weighed heavily on the institute's director. Lieutenant General Shinoda burst into tears as he spoke on 29 April, the emperor's birthday, of Japan's dire circumstances. For Kitazawa Ryuji, Shinoda's show of emotion that day underscored the gravity of the situation.

Corporations connected to Noborito also relocated production facilities to the relative security of the Japanese interior, many close to the institute's new branches. Among them were Nippon Wireless, Tokyo Shibaura Electric (Toshiba), and Japan High-Frequency. Noborito's new branches in the hinterland were thus able to continue, in the midst of severe supply shortages, their special projects and production for the decisive campaign in the home islands. Kitazawa Ryuji recalled that Fourth Section, responsible for producing equipment, had grown large in the final months. One of the section's products was a periscope disguised as a walking stick, a useful tool for commandos and guerrilla fighters to peer over and around walls in operations against Allied forces in Japan.[10]

Developing a Death Ray to the End

Unlike the successful development of Noborito's counterfeit bank notes, pathogens, and bombing balloons, First Section late in the war still had not fielded a death ray. Even so, the Japanese Army continued to fund the project. Military leaders knew that Japan's industrial base was far smaller than that of the United States and the British Empire. Japan had no chance to produce enough aircraft, warships, and other conventional arms to stave off defeat. An epochal technical breakthrough held out the only hope.

Colonel Satake explained the thinking behind pouring resources into such an unconventional project as a death ray. Yes, he admitted, there was no guarantee of success.

> However, from the standpoint that there would be no opportunity to change the outcome of the fighting without some extraordinary advance, a crucial decision was made to strengthen the research. With grim resolution, the plan was made to strive until the end to complete it, even if the enemy dared to come ashore in the Japanese mainland.[11]

While civilians trained with bamboo spears to defend against the coming Allied invasion, researchers of Noborito's First Section continued to work on the cutting edge of science for a wonder weapon. While some project personnel remained at Noborito, others withdrew to Nagano Prefecture's Kita-Azumi County, also known by its Japanese abbreviation, Hokuan. There they continued work from May at Noborito's new Hokuan Laboratory. Lieutenant General Shinoda remained at Noborito Branch while Major General Kusaba Sueki took charge at the new Hokuan Branch. He made his headquarters at Matsukawa National School, the primary school of the town by that name. Major Otsuki Toshiro, Major Takeda Teruhiko, and Dr Sasada Sukesaburo led over 100 men in research on various ultra-shortwave technologies. In addition to a death ray, researchers also worked to develop guided weapons.

Building some of the equipment for the project in the Hokuan Branch was a unit of the Japan High-Frequency Company, the Ikeda Laboratory, named for its location near the town of Ikeda. There the company manufactured magnetrons from May until the war's announced end in August. One device developed was an 80cm magnetron, manufactured under the supervision of Dr Ikebe Tsuneto according to the design of Professor Nishimaki Masao.

At war's end, there was taking shape a new vacuum tube factory for the Ikeda Laboratory, which planned to produce vacuum tubes with a 50cm wavelength and output of more than 100 kilowatts.[12]

By the war's end, Noborito researchers were working on a device in which, according to a postwar report of the US Navy, 'power taken from the magnetrons was fed into a parallel wire or coaxial transmission line, tuned externally, and from there the power was delivered to a dipole radiator near the focus of a parabolic mirror reflector.' At Hokuan, researchers were preparing experiments for converging the mirrors in a parabolic array 10 metres in diameter.[13] As part of a device with power output of 200 to 300 kilowatts, powered by an 80cm system, researchers were aiming to kill a rabbit at a distance of one kilometre.

The ultimate goal in the final months of the project, backed with a large budget of one million yen, was to kill enemy aircrew in flight. The project ended without killing any pilots, but some American prisoners almost died in the final experiments. Sasaki Tadashi, a future vice president of the electronics maker Sharp Corporation, recalled their peril:

> Imperial General Headquarters were pushing us to have the experiments succeed as soon as possible. There was even a plan to conduct experiments using American prisoners. The end of the war came just before the experiments were to be conducted. If we had actually carried out human experiments, we might have been to court-martialled. My first thought on hearing the emperor's broadcast announcing the war's end was that I had to survive without being taken before a court-martial. I remember tossing equipment for the experiments into Lake Suwa and getting away.[14]

Preparing for Decisive Battle with Poisons and Pathogens

Noborito had developed poisons and ways to hide some of them in food and drink, such as chocolate and whiskey. Second Section's probably was producing such products to poison the Allied forces that were soon to invade. The Nakano School, closely linked to Noborito, was responsible for organising civilians in guerrilla units. Nakano had put together a manual near the war's end that included a section on drug doses fatal to humans.[15] In addition to preparing to plant Noborito's poisons in enemy areas, Nakano officers could also have used Noborito data to write guidelines for poisoning with common drugs.

Researchers developing biological weapons for Noborito's Second Section took materials with them when they withdrew to more secure positions. In Fusan, the OGGC Livestock Hygiene Research Institute was near the coast. Heavy aerial mining of the waters around Fusan in the summer of 1945, part of the US Operation Starvation to block the shipment of food and other materials from ports in Chosen and China to Japan, threatened death and the destruction from errant parachute mines. Dr Nakamura Junji later recalled that it was the parachute mines that prompted the withdrawal of institute personnel to safer locations in the peninsula's interior. One department took its stores of 'highly virulent' rinderpest to an agricultural school in the interior city of Seishu in North Chusei Province. Other departments went to a separate site in South Chusei Province.[16]

Rinderpest from the OGGC Livestock Hygiene Research Institute was among the materials taken into the Japanese interior to counter the coming invasion of the Japanese home islands. Rinderpest, other pathogenic materials, poisoned food, and 16mm film documenting Second Section's projects were taken to Noborito's Nakazawa Branch in Nagano Prefecture. Noborito's materials were to be part of a campaign, Major Ban Shigeo explained after the war, in which the Japanese Army would infect wells and other water supplies with pathogenic and toxic materials to sicken and kill the enemy. Noborito pens filled with bacteria for poisoning water supplies had already gone with operatives of the Nakano School on their missions to Japanese islands of the Ryukyu chain and elsewhere. In the decisive fighting in defence of the Empire, expected to take place in the interior of the main island of Honshu in 1946 following an invasion of Kyushu in late 1945, Japanese leaders at Matsushiro Underground Imperial General Headquarters, their soldiers in the field, and civilians fighting as militia, for all Japanese were expected to fight, were to remain healthy by drinking water filtered through water filters developed by the Japanese Army's major BW organisation, Unit 731.[17]

Making Balloons for a 'Second Season'

At the Yamaguchi Higher Girls' School, Inoue Toshiko and her classmates spent the final days of the war in the summer of 1945 making balloons and drilling with bamboo spears. The announced goal was for each student to kill one American soldier before dying. The balloons were for killing Americans across the Pacific. Inoue, who had started making balloons at her school on 1 August 1944, only stopped on 15 August 1945, when the emperor's radio broadcast brought the programme to an end.[18]

In Hsinking, capital of Manchukuo, Sakiyama Hiromi, 15 years old, had been labouring since June with classmates under military supervision at the Hsinking Shikishima Higher Girls' School to produce bombing balloons. She and her classmates continued working until armed forces of the Soviet Union invaded Manchukuo on 9 August. She reported for work that morning after Soviet aircraft had bombed the capital, only to find the Japanese soldiers stationed at the school preparing to retreat south as Soviet forces advanced from the north. The girls were dismissed from work, left with their families to fend for themselves as the Red Army overran Manchukuo.

At Takasaki Higher Girls' School in Gunma Prefecture, Murata Kiyoko was working with classmates to make balloons and self-destruct explosive charges for them. The students continued working at their school-turned-factory until Takasaki suffered an air raid on 14 August. The next day, as instructed, the girls gathered in the schoolyard to listen at noon to an important radio broadcast. It was the voice of the emperor, announcing the war's end. While the students stood assembled there, kamikaze pilots from a nearby airbase flew low above the school, scattering leaflets proclaiming that Japan had not lost. After hearing the broadcast, the girls started sobbing, leaning upon one another for support. Murata was unsure exactly why they crying. Perhaps it was a sense of shock, having heard repeatedly from all the adults around them that Japan would never be defeated.

That young women in Japan were continuing to make bombing balloons until the end strongly suggests that the Japanese Army planned to start, in the words of Noborito historian Yamada Akira, a 'second season' of bombing from November 1945, when the jet stream would again be available and when Allied forces were set to invade the main island of Kyushu. Meanwhile, in Manchukuo, according to Sakiyama, she and her classmates had been constructing balloons five metres in diameter, only half the size of those built to attack the United States. The reason, according to Noborito expert Watanabe Kenji, is that Japan was planning to use the smaller balloons to attack the Soviet Union. Furthermore, Watanabe has speculated, the Japanese Army intended to use biological weapons prepared by Unit 731 as the payload.

Yamada Akira has suggested the possibility that the Japanese Army had reversed its decision in July 1944 not to load biological weapons on the bombing balloons in light of the coming decisive battle in the Japanese home islands. Unit 731's insects and pathogens would likely survive the much shorter trip at lower altitude to targets in the Soviet Far East.

Furthermore, Noborito late in the war was working on a larger balloon. Major General Kusaba, the lead for the Fu-go project as director of Noborito's First Section, was aiming to develop a new type of bombing balloon. Fifteen metres in diameter, the larger balloon would be capable of rising to 14 kilometres, an altitude at which stronger winds would make possible attacks against the United States even in summer. Testing had only just began and only two of the larger balloons had been made when the war ended.[19]

The 1949 trials in the Soviet Far East city of Khabarovsk of IJA officers associated with the Kwantung Army's BW programme further point to Japanese plans for biological warfare. General Yamada Otozo, commander-in-chief of the Kwantung Army, pleaded guilty 'to having exercised direct guidance of preparations for conducting biological warfare against the USSR, China, the Mongolian People's Republic, England, the USA, and other countries.' He stated that Japan 'could have' used biological weapons against 'the USA and other countries' had the Soviet Union not gone to war with Japan in August 1945. Lieutenant General Kajitsuka Ryuji, director of the Kwantung Army's medical section, also admitted that Japan had prepared to conduct biological warfare 'chiefly' against the Soviet Union and the Mongolian People's Republic, as well as against the United States and Great Britain.[20]

Countering Biological Weapons on Bombing Balloons

American military authorities by the summer of 1945 saw relatively little threat in the conventional payloads of the bombing balloons but expected that the paper intercontinental bombers would arrive in the coming months carrying biological weapons. The 15kg explosives released by balloons floating freely on the winds and without targeting mechanisms over the vast expanse of North America could not devastate cities or military targets the way that bombers or rockets could do. Japanese incendiaries released from the balloons were capable of igniting forest fires in the American West's drier months. Even so, a special committee of the Office of Scientific Research and Development (OSRD) concluded in a report of June 1945 that the damage inflicted by the bombing balloons, absent a change in their payload, would be less costly than the steps required to defend against them: 'Only if extensive resources are diverted to defensive measures will the balloon incendiary campaign be of profit to the enemy.'[21]

Much more worrisome was the threat of biological weapons delivered by bombing balloons. The War Department's Military Intelligence Division in December 1944 tasked the chief of the MIS to 'determine as soon as possible' if the bombing balloons could introduce 'bacteriological warfare materials' into the United States. MIS in January 1945 organised a team to investigate the bombing balloons, including a scientific branch with instructions to 'check bacteriological warfare possibilities: hazards to humans, animals, and plant life.' In February 1945, the CWS, working with the Surgeon General's Office, took responsibility for studying the BW problem. While the bombing balloon's components continued going to the Technical Air Intelligence Center in the District of Columbia, 'BW specimens' were to go 'under guard or by qualified courier' to the CWS.[22]

The bombing balloons as weapons made little sense other than as a way to wage biological warfare. A report of the Army Service Forces in February 1945 to the commanding general of the Seventh Service Command made the following assessment:

> Present evaluation is that the pay load carried by these balloons is such as to indicate that only a relatively small amount of incendiaries or explosives can be carried by any one balloon. The carrying of human cargo although possible is highly improbable. A sufficient amount of disease-producing agencies can be carried by each balloon to warrant that special attention be given this activity.

Japan had not only launched against the United States bombing balloons seemingly made for carrying BW payloads but had already waged biological warfare. The Japanese Army by this time had conducted BW attacks by aircraft against targets in central China, dropping in 1940 infected materials on areas of Chekiang Province, which resulted in deadly plague outbreaks. Chinese health officials confirmed sudden eruptions of plague cases in areas over which witnesses had earlier spotted aircraft dropping infected grain and fleas. US military authorities by 1945 had also received intelligence on Unit 1644's BW activities in Nanking, reportedly involving the development of BW materials for delivery via bombing balloons. Washington also had intelligence on the development by Unit 731 in Manchukuo of bombs for cholera, diphtheria, and typhoid.[23]

Moreover, intelligence derived from such sources as captured documents and interrogated prisoners of war suggested that the Japanese Army would be launching thousands of balloons in the autumn of 1945. The Japanese Army had been using balloon radiosondes to transmit data on the jet stream and balloon performance. Close to half the signals that the United States detected from those radiosondes were noted after mid-April. The absence of bombing balloons spotted over the United States after that month seemed only a pause before new ones – perhaps this time carrying deadly biological payloads – started arriving in the autumn when the jet stream resumed its rapid flow from west to east. Another concern was that, as time passed from the first balloon recovery, officials had noticed an improvement in manufacturing quality – produced first by hand, then by machine in some cases – suggesting that the first balloons had been sent as a test. Dr Lincoln LaPaz, technical director of the Second Air Force Operations Analysis Section, was among the scientists who 'believed that the next step on the Japanese war plan, scheduled for the autumn of 1945, was to be a balloon-borne bacteriological attack which almost certainly would have hurt us greatly.'[24]

In the spring of 1945, a tragic explosion in Oregon strengthened earlier confirmation for the Japanese Army that their bombing balloons had been reaching the mainland United States. After initial press reporting of an early discovery in November 1944, US authorities had persuaded American media to refrain from further reports in order to keep from the Japanese evidence that their weapons were reaching North America. The blanket censorship ended in the spring of 1945, following the tragic death by balloon bomb on 5 May of a woman and five children. The Reverend Archie Mitchell had taken his wife and five children from his church on a picnic to a wooded area near Bly, Oregon. Mrs Mitchell and the children died in the blast of an unexploded bomb from the payload of a balloon that had landed there weeks before. They were the only American casualties of Noborito's bombing balloons.

To alert the public and avoid further deaths, the US government ended after the incident its policy of blanket censorship. On 22 May, the Army and Navy went public with the news that Japanese bombing balloons were reaching the continental United States. On 31 May, Secretary of War Robert Patterson disclosed to the press some of the details on the deaths in Oregon. However, a memo of the Joint Security Control, a joint body of Army and Navy Intelligence under the Joint Chiefs of Staff, to Byron Price, director of the Office of Censorship, called on him to send a confidential memorandum to all editors requesting that the press continue to refrain from publishing photographs of recovered bombing balloons. The concern was that pictures of the balloons, which Noborito had designed to self-destruct and leave no trace, would 'inform the Japanese what parts of the mechanism are malfunctioning and thus assist them to correct the design to our detriment.'[25] Nevertheless, the Japanese Army now had clear evidence after months of news blackout that their new weapon worked.

Noborito's biological weapons could possibly have caused widespread disease and death for people, livestock, and crops. The US military had essentially failed to stop the balloons. Authorities estimated on the basis of close to 'fewer than 300' sightings that approximately 900 balloons, nearly one in 10 of the approximately 9,300 launched, had reached North America. Most had dropped their payloads or come to earth in the American West, but some had floated as far east as Detroit in the Upper Midwest.

Aircraft had managed to shoot down only two balloons over the continental United States and a few more over the Aleutian Islands. The major problem, as Dr LaPaz put it, was that the 'crude radar of 1945' had little chance of detecting a weapon with so little metal in it, leaving the United States no countermeasure 'except to fight the bacteria when they came.' Indeed, a joint Army-Navy defence plan issued on 15 August, the day that Tokyo broadcast the Japanese decision to end the war, indicated the absence of an effective radar defence and spelled out responsibilities for 'defence against biological warfare in connection with Japanese free balloons' in the western United States. The problem, as the report noted, was that local authorities would have been in the dark until 'an uncommon or unseasonable disease or unexplained epidemic develops in humans, animals or vegetation in a given area.'[26]

Ending the War, Destroying the Evidence

In advance of the emperor's noon radio broadcast of 15 August there came that morning secret orders from the the Military Affairs Section of the Army's

Ministry's Military Bureau to Noborito for the complete destruction of all sensitive documents and secret weapons. Institute personnel immediately went to work to remove every trace of incriminating evidence.

When the war ended, Noborito researchers were developing a large magnetron comprising multiple segments and output of 100 kilowatts for 400cm continuous waves (CW). The development team had also planned to link 10 of the magnetrons in parallel for a CW output of 1,000 kilowatts. It was not to be. Following orders, team members threw the partially completed magnetron into Lake Kizaki, near the headquarters of Hokuan Branch at Matsukawa.[27]

At the agricultural school in Seishu, where researchers of the OGGC Livestock Hygiene Research Institute had withdrawn from Fusan, researchers burned sensitive equipment and materials. At the Noborito Branch, Second Section's Group 7, which under Kuba Noboru had directed the rinderpest programme in Fusan and worked on other BW projects, received orders to disguise his group as a sterilisation station and burn everything that did not fit that story. Group 7 had transported stocks of powdered rinderpest, as well as documents, and 16mm film of its activities, to Nagano Prefecture. All those incriminating materials, biological weapons and explosives among them, were piled two days after the Emperor's broadcast into a pit of approximately 20 metres in diameter in the town square in Nakazawa, then set ablaze. The resulting blast blew out the glass from windows in surrounding buildings.[28]

Avoiding prison or a death sentence at the hands of the war's victors was a key point in all this destruction. Concern that their superiors would be branded war criminals, according to a first lieutenant assigned to Second Section, motivated personnel to burn the documents, equipment, and materials.[29] No doubt the lieutenant and other junior personnel were thinking as well about could happen to themselves as well in the event that the Allies unearthed incriminating evidence.

While Noborito destroyed its biological weapons, some of the institute's poisons escaped the destruction. Many military officers, in despair on hearing the Emperor's radio broadcast and anxious about their fate in the coming Allied occupation, decided to kill themselves. Noborito Fourth Section's Kitazawa Ryuji, responsible for the institute's stock of acetone cyanohydrin and other deadly materials, received at that time a visitor's request on behalf of officers from the War Ministry, General Staff, and the Kempeitai for doses of Noborito poison as a way to kill themselves. Word that Noborito had stocks of poison had spread

Above: Colonel Shinoda Ryo, center of first row, in uniform with sword, at the Army Scientific Research Institute (ASRI) in May 1939 with subordinates in Second Department's Group 4. The photograph was taken to mark the group's move from Tokyo to Noborito. (Courtesy of the Defunct Imperial Japanese Army Noborito Laboratory Museum for Education in Peace (1757))

Left: Inset of Colonel Shinoda

Below: Japanese Army certificate of award for technical merit, dated 14 April 1943 and presented in the name of Army Minister Tojo Hideki, at that time also prime minister, to Noborito's Major General Shinoda Ryo and Technical Captain Ban Shigeru. (Courtesy of the Defunct Imperial Japanese Army Noborito Laboratory Museum for Education in Peace (13))

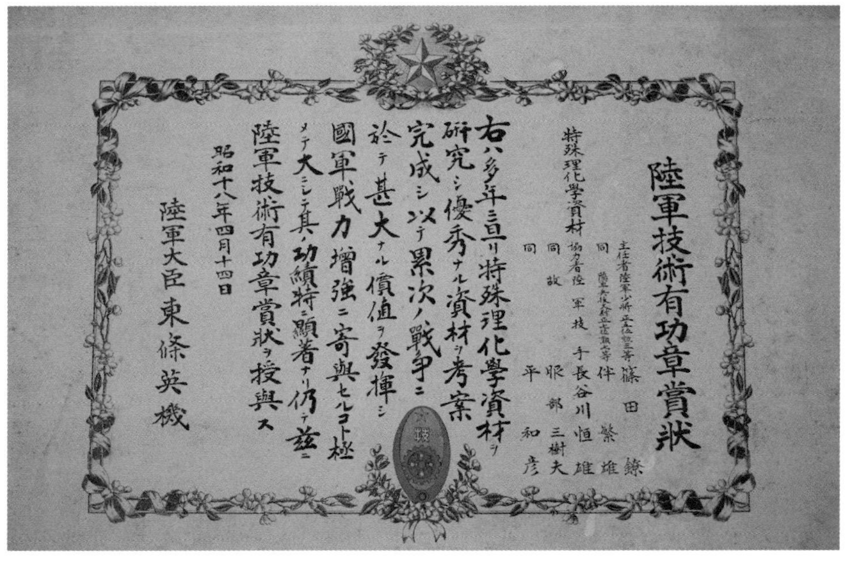

陸軍技術有功章賞狀

特殊理化學資材

主任者陸軍少將 [illegible] 篠田鐐

同 [illegible] 伴繁雄

協力者陸軍技手 長谷川恒雄

同 同 服部三樹夫

同 故 平和彦

右ハ多年ニ亘リ特殊理化學資材ヲ研究シ優秀ナル資材ヲ考案完成シ以テ累次ノ戰爭ニ於テ甚大ナル價値ヲ發揮シ國軍戰力增強ニ寄與セルコト極メテ大ニシテ其ノ功績特ニ顯著ナリ仍テ茲ニ陸軍技術有功章賞狀ヲ授與ス

昭和十八年四月十四日

陸軍大臣 東條英機

Above left: The American former military intelligence officer Herbert Yardley shocked Japan with his 1931 book, a best seller in Japanese translation, which revealed that Washington had negotiated a naval treaty with Tokyo while accessing intelligence gained from Yardley's breaking of Japanese diplomatic codes. At the ASRI, and later Noborito, researchers read with keen interest the book's passages on secret inks and other intelligence tools. (Courtesy of the US National Archives (111-SC-55452a))

Above right: Stone monument, erected at Noborito in 1943 with Shinoda and Ban's award money, to comfort the souls of animals sacrificed in various projects. (Author's photograph)

Major Prince Mikasa, first row, center, poses for a photograph on his visit to Noborito in October 1944. Seated to the right is Lieutenant General Shinoda. The visit of Emperor Hirohito's younger brother points to the high expectations held for Noborito. (Courtesy of the Defunct Imperial Japanese Army Noborito Laboratory Museum for Education in Peace (514))

Postwar photograph of balloon launch site at Otsu, Ibaraki Prefecture. (Courtesy of the US National Archives (SC 284816))

Above left: US military personnel in postwar inspection of Otsu launch site, standing before storage tanks of vanished hydrogen plant. (Courtesy of the US National Archives (SC 284185))

Above right: A bombing balloon, recovered at Alturas, California, on 10 January 1945, is inflated for testing at a military base in California. (Courtesy of the US National Archives (SC-226135))

Above left: US military personnel handling a recovered bombing balloon's mechanism – including, from top to bottom, its battery case, barometer box, activating fuse, and ballast ring – at Naval Air Station Moffett Field, California, May 1945. (Courtesy of the US National Archives (80-G-326355))

Above right: A high-explosive shell recovered from a bombing balloon, Naval Air Station Moffett Field, California, March 1945. (Courtesy of the US National Archives (80-G-326345))

Counterfeit 10 yuan notes of China's Bank of Communications, printed at Noborito. (Courtesy of the Defunct Imperial Japanese Army Noborito Laboratory Museum for Education in Peace (61))

Above left: Bust of Subhas Chandra Bose at Tokyo's Renkoji Temple, where the remains of the Indian National Army's leader lie. Major General Iwakuro Hideo, who oversaw operations against British India, had Noborito produce counterfeit rupees for them. He and Lieutenant General Arisue Seizo, the Japanese Army's last intelligence chief, were members of Japan's postwar Subhas Chandra Bose Academy, which erected the bust to Bose's memory. (Author's photograph)

Above right: Major Ban Shigeo, in his later years, standing before a collection of water filter cartridges intended for use in Japan's planned decisive military campaign in the Japanese main islands, expected to start in late 1945. (Courtesy of the Defunct Imperial Japanese Army Noborito Laboratory Museum for Education in Peace (331-11))

Lieutenant General Kawabe Torashiro, second from left, with Major General Charles Willoughby, at left, in Manila, 19 August 1945, before the formal surrender, to discuss details of Japan's military occupation. Kawabe would become an important intelligence broker in postwar Japan. US Army Signal Corps photograph. (Courtesy of the Harry S. Truman Library (SC 2014-3335))

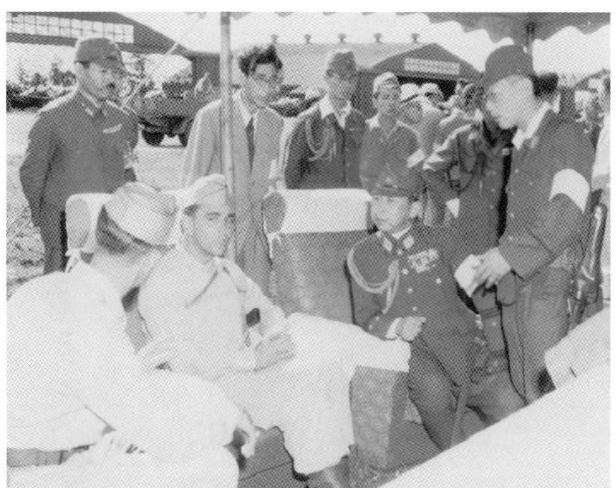

Above left: Lieutenant General Arisue Seizo, in uniform, with ceremonial sword and the shoulder aiguillette of a staff officer, discusses details of Japan's occupation with Colonel Charles Tench, head of General Douglas MacArthur's advance team, at Atsugi airfield, near Tokyo, 28 August 1945. Arisue, too, would become a key intelligence broker in postwar Japan. (Courtesy of the US National Archives (SC 210609-S))

Above right: Two soldiers of the US Eighth Army in the Republic of Korea in 1954 model military uniforms and weapons of the Democratic People's Republic of Korea and the People's Republic of China. The US military maintained stores of foreign uniforms and weapons for missions behind enemy lines. Noborito veterans working for the US government forged during and after the Korean War documents for operatives to take on such missions. (Courtesy of the US National Archives (SC 453647-W))

Former Noborito employees pose for a group picture in January 1946 at their first New Year's gathering after the war. (Courtesy of the Defunct Imperial Japanese Army Noborito Laboratory Museum for Education in Peace (1394))

Above left: Meiji University's Building 5, as it appeared in 1995, had housed a wartime printing plant for Noborito's counterfeiting programme. University authorities razed the building in 2011. (Author's photograph)

Above right: Building 26, as it appeared in 1995, is thought to have served as a storehouse for Noborito's counterfeit bank notes. The building was razed in 2009. (Author's photograph)

Above left: Front of the stone monument placed by surviving veterans in 1989 on the grounds of what was Noborito's Yagokoro Shrine. (Author's photograph)

Above right: A traditional *torii* gate marks the entrance to the former Yagokoro Shrine, known today as the Ikuta Shrine. Visible at right is the stone monument erected by Noborito veterans in 1989. (Author's photograph)

Front of the Defunct Imperial Japanese Army Noborito Laboratory Museum for Education in Peace. Known after the war as Building 36, the structure housed wartime laboratories where Noborito's Second Section developed biological weapons. It is Noborito's only surviving building. (Author's photograph)

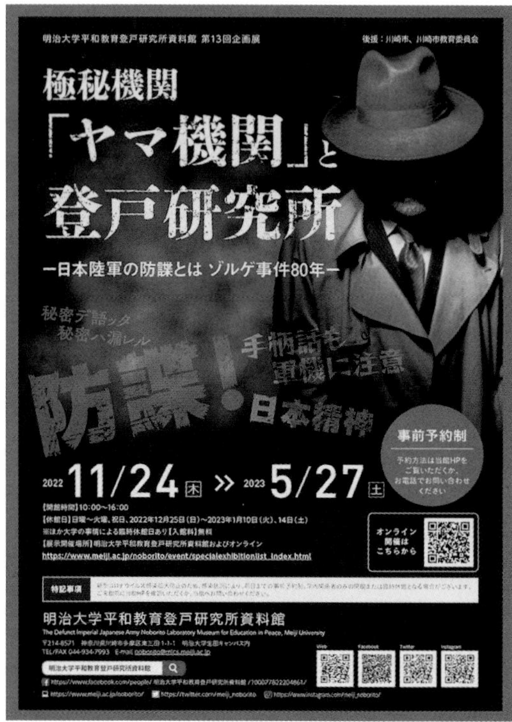

A poster advertising the museum's exhibition on the connection between the Yama Agency – the Japanese Army's secret counterintelligence organ – and the Noborito Research Institute. The exhibition opened in 2022, the 80th anniversary of the public revelation of the Soviet spy Richard Sorge's arrest in Tokyo. Yama and Noborito played parts in the breaking of Sorge's spy ring. The museum organises various conferences and exhibitions to bring to the public information on Noborito's history. (Courtesy of the Defunct Imperial Japanese Army Noborito Laboratory Museum for Education in Peace)

throughout the military high command in Tokyo via 8th Section of AGS Second Bureau. Kitazawa recalled years later that he had thus handed over three cases of the poison, each case holding some 30 to 50 doses. Kitazawa had passed some of the poison via an authorised 'official route,' but he would not discount the possibility that other ampoules of poison had been smuggled out.[30]

Noborito's poisons almost killed some children in one incident. A Japanese soldier, spotting some chocolate in a burn pit at a school in the village of Ina, in Nagano Prefecture, offered the treat to some children who were helping with the destruction of materials. The soldier was unaware that he had handed out poisoned chocolate developed in Major Hijikata's group and brought from Noborito. Luckily for the children, a Noborito officer caught the mistake. The children had their stomachs pumped in time to save their lives.[31]

Some of Noborito's counterfeiters had withdrawn from the Tokyo area in the war's final months. As the bombs fell week after week around the capital, Third Section's North Group sustained slight bombing damage. Due to the particular requirements for paper and printing facilities, the decision was made in March 1945 to move part of Third Section to an agricultural warehouse in Fukui Prefecture, near the Sea of Japan. The deciding factor was the location of the factory of the Kato Paper Company. Third Section would be able to use much of the factory's equipment for its operations, which would save considerable effort and resources in the relocation. When the war came to an end, Third Section passed printing machines to the government's Printing Bureau. Section personnel dumped into the Sea of Japan equipment that was incriminating or of no civilian use. As for the counterfeit bank notes and other materials, Third Section's Oshima recalled that it took a week to burn it all. Colonel Yamamoto, seeing all his efforts as Third Section's director go up in flames day after day, the rising smoke disappearing into the summer sky, thought in anguish: 'It was all in vain!'[32]

Destroying Evidence of the Most Sensitive of Secrets

The document recording the order for the destruction of all incriminating evidence of research into biological warfare came with the bland title *Special Research Disposal Procedures*. Issued by Military Affairs Section of the Military Affairs Bureau and dated 15 August 1945, the day of the Emperor's noon broadcast announcing to the Japanese people the war's end, the document's first section included the order: 'Take measures immediately to destroy all evidence of special research that would be to our disadvantage if the enemy were to obtain evidence of it.'

Listed first were Noborito's Fu-go project and the Noborito Research Institute as an organisation. Second came the Kwantung Army and its field units for BW research and development: Unit 731 and Unit 100. Third was the Army First Provisions Main Depot, an organisation that had been working on the remote island of Tanegashima to develop grain smut (smut fungus) as a weapon for Noborito's bombing balloons to drop on American crops. Listed fourth and fifth were special research activities related to medical and veterinary science.[33]

The author of the document, Lieutenant Colonel Niizuma Seiichi, wrote of having first contacted Lieutenant Colonel Kusakari Michitomo at the AOAH at 08:30 to pass the order regarding the Fu-go project and Noborito. This raises the possibility that the bombing balloons and other Noborito projects were more sensitive than even the horrific human and animal experiments conducted by Unit 731 and Unit 100. The two Kwantung Army BW field research organisations had carried out cruel vivisection and other medical experiments on Chinese and Russians prisoners. Noborito, too, as Major Ban admitted, had tested poisons on prisoners in China.

That the Fu-go project and Noborito were at the top of the list, above the Kwantung Army's Unit 731 and Unit 100, suggests the possibility that Noborito researchers developed for delivery by bombing balloon pathogens for killing people as well as those for destroying livestock and crops. Noborito's BW projects, even if no more horrible than those of Unit 731 and Unit 100, were those most clearly developed to attack the United States. The Japanese Army would have nearly two weeks to destroy incriminating evidence of special research at Noborito and elsewhere before the advance elements of the Allied Forces under the command of General MacArthur arrived to occupy Japan.[34]

CHAPTER 7

DEALING WITH THE VICTORS IN OCCUPIED JAPAN AND KOREA

> I was certainly close to Mussolini. I liked and respected him. However, I am by no means a fascist. I am an honourable Japanese soldier. In particular, I have associated over many years with foreigners. Consequently, I have no complex whatsoever in regard to them, especially Westerners.
>
> – Lieutenant General Arisue Seizo, chief, Second Bureau, Army General Staff[1]

Victors Seeking Spoils of War

With the victory of the Allied powers over Germany and Japan in 1945, the victors were quick to seek the spoils of war. The United States, Soviet Union, Great Britain, France, and China went searching for scientists, engineers, and intelligence in areas of military science and technology where Germany and Japan excelled.

In Germany, the victors were eager to acquire aerospace scientists and engineers. Washington brought that year Wernher von Braun, a leading developer of Germany's V-2, the world's first long-range ballistic missile, to the United States, where he was one of more than 100 V-2 veterans who worked to develop US missiles and rockets. Moscow took over 200 German aerospace experts to the Soviet Union. Paris enlisted a number of V-2 engineers among the more than 1,000 German researchers recruited to France between 1945 and 1950. A research director of the aircraft company Junkers, located by French intelligence officers in Vienna took French citizenship and became a designer of Mirage fighter aircraft.[2]

In Japan, BW technology was the top prize. According to a top-secret OSS memorandum of 27 September 1945 on biological warfare,

'The Japanese had perhaps the best informed scientists in BW investigations of any nation in the world and as such were more of a potential threat in that direction than Germany.' A postwar CWS history assessed that, 'Japanese activities were better organised and more comprehensive than those of Germany. Japan appears to have started biological warfare studies as early as 1936, with the principal wartime research centred in a Defence Intelligence Institute near Harbin in Manchuria, where 2,500 people were employed at the peak of operations.'[3]

In East Asia, Washington and Moscow acquired spoils of war in areas under their control. Washington summoned military officers, scientists, and technicians for interrogation, gathered data, and collected equipment in its occupied areas: Japan and Korea south of the 38th parallel. The Soviet Union sought treasure in its own areas: Manchuria and northern Korea. Lieutenant General Ishii Shiro and other BW experts in areas under American control exchanged data for immunity from prosecution for war crimes. Japanese fared less well with the Soviets. Moscow put on trial in December 1949 in Khabarovsk a dozen officers of the Kwantung Army involved in biological warfare, handing down prison sentences of between two to twenty-five years.[4]

Moscow's wartime allies were also quick to enlist military intelligence officers of Germany and Japan for their Soviet expertise. In Germany, Colonel Reinhard Gehlen, who had been chief of intelligence for the German Army's Foreign Armies East (FHO), placed before the Americans hidden crates of military intelligence on the Soviet Union. In Japan, Lieutenant General Arisue Seizo, the Japanese Army's last intelligence chief, also unearthed cached trunks of Japanese military documents on the Soviet Union to come to terms with his new partners. Washington accepted the intelligence as the wartime alliance with the Soviet Union soon gave way to conflict. War seemed on the horizon. The Chinese Nationalists were also active, enlisting Japanese military code breakers and other intelligence veterans – employing some from among those in China at the war's end and recruiting others from Occupied Japan – for work against the Soviet Union as well as the Chinese Communists.[5]

Moscow dealt harshly with Japanese military intelligence officers taken prisoner at war's end in Manchuria and Korea. Most of the hundreds of thousands of Japanese who survived the ordeal of forced labour in Siberian camps returned to Japan within several years, but more than 200 IJA intelligence officers remained in Soviet detention until their release in a period of thaw in Japanese-Soviet relations in the mid-1950s. Only in 1956 did the last intelligence veterans return to Japan, including more than 50 officers of the Nakano School who sailed home from Siberia on the final repatriation ship.[6]

From Mussolini to MacArthur

Lieutenant General Arisue, who would play a pivotal postwar role in brokering Japanese intelligence to the United States, was a remarkable military man. A member of IMA Class 29 (1917) and Army War College Class 36 (1924), Arisue was an air officer who spent much of his career in foreign and intelligence affairs. The War Ministry sent him in 1928 to Italy for three years of language and military studies at the War College in Turin. He directed for a year the foreign affairs section in the Military Affairs Section of the Military Affairs Bureau before returning to Italy for a three-year tour in 1936 as military attache to the Japanese embassy. In August 1942, Arisue became head of military intelligence as director of AGS Second Bureau.

In August 1945, Arisue volunteered to fly to Manila to arrange details of Japan's surrender and occupation with the staff of General Douglas MacArthur, who was to oversee Occupied Japan as Supreme Commander for the Allied Powers (SCAP). Instead, Lieutenant General Kawabe Torashiro, deputy chief of the Army General Staff, an officer whose career included tours as military attache to the Japanese embassies in Moscow and Berlin, led the delegation. Kawabe was an inspired choice. He established immediate rapport with MacArthur's intelligence chief, Major General Charles Willoughby. Kawabe, seated next to the German-born Willoughby in the back of his car on the ride from the airfield, declared that he would prefer to speak to him in German. Willoughby showed his desire for close working relations with Kawabe, and by extension the Japanese Army, by leaving in his room that evening a gift of cigarettes and whiskey together with his calling card.

Determined to play a leading role, Arisue stepped forwards to serve as liaison to MacArthur's advance team, which flew into Atsugi air field on 28 August. Arisue had brushed aside the concern of Prime Minster Prince Higashikuni Naruhiko that his past association with the Italian dictator Benito Mussolini made him unsuited for the task, insisting that he was no fascist and, more importantly, that he knew how to handle

Westerners.[7] Arisue succeeded in conducting the Atsugi meeting without incident, then established a liaison office near the Yokohama Grand Hotel, MacArthur's temporary headquarters.

On 31 August, the day after MacArthur had flown into Japan, Arisue received from the hotel a telephone call from Colonel Frederick P. Munson, whom he had met years earlier when the American had served a tour as a language student in Japan. Now chief of the Japan liaison section in SCAP's G-2, Munson invited Arisue to a meeting at the hotel. The following week, only a few days after the Japanese surrender ceremony of 2 September on board the battleship USS *Missouri*, Munson introduced Arisue to Willoughby. When MacArthur left Yokohama to establish SCAP General Headquarters (GHQ/SCAP) in the imposing Dai Ichi Building, situated across the moat from the Imperial Palace, Arisue moved his organisation into the nearby Japan Club, from which he could see the offices of Willoughby and Munson.[8]

In Occupied Japan, GHQ/SCAP worked with Japanese in the former Army and Navy ministries, which the victors turned into the First (Army) and Second (Navy) Demobilisation bureaus. Arisue became a key point of contact at the First Demobilisation Bureau, using the subordinates in his Arisue Agency to arrange for Japanese to appear before Allied officials for questioning. Arisue, once close to Mussolini, bonded with Willoughby and other GHQ/SCAP officers. An application of 20 April 1947 from the First Demobilisation Bureau to retain Arisue at his post as a 'non-regular official' attached to its Document Section, despite a directive for his purge from office, described him as 'in charge of the liaison business between the Allied Forces and the 1st Demob. Bureau.' Further, 'He has also may friends and acquaintances in the GHQ. For this reason, he is the most suitable person in carrying out mutual liaison business smoothly.' Arisue ran his agency from August 1945 to March 1946, operated as an official of the First Demobilisation Bureau from December 1945, and worked as an advisor to the US military in Japan from July 1946 to December 1956.[9]

Intelligence officers on both sides enjoyed good relations from early in the occupation. On 24 September, Arisue and Kawabe called at the Imperial Hotel on Willoughby and his subordinate, Colonel Sydney Mashbir, who was fluent in Japanese. Willoughby and Kawabe conversed in German, while Mashbir spoke to both visitors in Japanese and interpreted for his boss. Willoughby expressed his sympathy to his Japanese counterparts for their defeat. He was seeking to develop rapport with the two Japanese officers, but he also spoke from the heart. Before the war, Willoughby had written of Asia's choice in politics as either 'red sickle or rising sun.' He favoured the latter. Arisue and Willoughby spoke to one another as fellow intelligence officers on various topics that night and on later occasions. Arisue recalled commiserating with Willoughby on the disagreements common in relations between military officers in operations and those in intelligence, and those between military and naval officers.[10]

Arisue wrote in one of his memoirs of learning that US officials would request interviews with officers, soldiers, and civilians for three reasons: (1) to collect information to try Japanese suspected of war crimes, (2) to prepare war histories, and (3) to gather intelligence 'against the Soviet Union, a future hypothetical enemy of the US military in East Asia.'[11] Arisue would manage the requests, producing individuals for the Americans to question while keeping incriminating intelligence hidden.

Also keeping such intelligence from Allied eyes was Lieutenant Colonel Niizuma Seiichi. The very officer who had ordered at war's end the destruction of all incriminating evidence of IJA 'special research', he worked after the war to keep military secrets and Japanese officials safe. Niizuma in the war had overseen special research projects from the Military Affairs Section of the War Ministry's Military Affairs Bureau. In Occupied Japan, he managed requests for technical intelligence.

Arisue, Niizuma, and other Japanese military officers worked in particular to keep incriminating evidence of biological warfare from the victors. Lieutenant Colonel Murray Sanders, a medical officer from Camp Detrick, the centre of BW research in the United States, was responsible for investigating Japanese biological warfare as a member of the scientific intelligence survey team led by Karl Compton, physicist and president of the Massachusetts Institute of Technology (MIT). On 1 October, Sanders met Niizuma in the Dai Ichi Building at the start of his investigation. Interpreting for Niizuma was Lieutenant Colonel Naito Ryoichi, a medical officer and key member of the Japanese Army's BW research and development network. Murray was unaware that Niizuma had ordered the destruction of incriminating evidence on special research projects. Indeed, he remained ignorant for a time even of Naito's true identity. Murray left Japan several weeks later without ever learning the details of the IJA's work to develop offensive BW capabilities.

Murray's successor, Lieutenant Colonel Arvo T. Thompson, also failed to uncover intelligence on offensive biological warfare, including the IJA's experiments on human beings and the field tests in military operations in China. Thompson learnt little in questioning Lieutenant General Kitano Masaji, a medical officer and successor to Ishii Shiro as commander of Unit 731, after Kitano returned in 1946 to Japan from China. Kitano had met Arisue before his first meeting with Thompson and almost certainly worked out in advance a script to recite to the Americans, as others had done. Early in the Occupation, a memorandum noted that MIT's Dr Compton had been 'inclined to doubt' Japanese denials that they had engaged in offensive biological warfare' and that his group had overheard Japanese saying among themselves that they were 'not supposed to talk' about it. It was only when the Japanese later received credible assurances of immunity from prosecution that investigators obtained data and details of the military's offensive BW activities.[12]

Aiming for Intelligence Targets

The US Army by the end of the war had compiled a trove of intelligence, its Japanese files arrayed in offices from the Washington Document Center in Camp Ritchie, Maryland, to the Counter Intelligence Corps (CIC) Library in liberated Manila. In May 1943 the ATIS, under the command of Colonel Mashbir, had produced the document *Alphabetical List of Japanese Army Officers*. The ATIS report was a translation of information from the Japanese Army's *Register of Officers* and its *Register of Reserve Officers on the Active List*, issued the previous year.

The document listed Shinoda Ryo, a major general in 1942, as an engineer officer and 'Chief 9 Research Inst Army Tech HQ; Attd. GS HQ.' MacArthur's intelligence organisation thus knew before the Occupation that Shinoda commanded the 9th research institute under the ATH, predecessor to the AOAH and its 9th ATRI. The document also identified Shinoda's chief subordinates – the directors of Noborito's First, Second, and Third Sections, Kusaba Sueki, Yamada Sakura, and Yamamoto Kenzo – as officers assigned to an unspecified ATH institute.

The document listed even Noborito group chiefs, including Major Ban Shigeo. This and other intelligence, gave investigators some idea where to start. But the US Army still lacked details on Noborito. Military Research Bulletin No. 21, 'Japanese Intelligence,' issued on 15 August 1945 by the MIS in the War Department's Military Intelligence Division

(MID), listed the AOAH's ten ATRIs. The document included the general areas of activity for each institute, such as ordnance technology and explosives for the 8th ATRI. The purpose of the 9th ATRI (Noborito), however, was 'not known.'[13] The US Army knew at least as early as 1943 that Shinoda Ryo headed the institute but was seemingly unaware of what he was directing. With such intelligence, Compton's group arrived in Japan with a prepared list of Japanese scientists.[14]

Intelligence and technology were priority targets in Operation Blacklist for the occupation of Japan. A document outlining the planned conduct of 'Blacklist' operations, classified top secret, called for the 'immediate seizure of Intelligence Objectives' and the surveying of 'Japanese Military, scientific and technical developments and their application to peace and war.' With a nod to Japan's 'considerable progress in their industrial economy, science and technology since the beginning of the war,' the document section 'Economic, Scientific and Technological Intelligence' included the instruction that those advances 'should be fully investigated and exploited.'[15]

However, Major General Willoughby cautioned Lieutenant General Richard Sutherland, General MacArthur's chief of staff, in a memo of 27 August, soon before MacArthur's landing in Japan, against taking 'punitive measures.' Willoughby had reached an understanding with Kawabe's delegation in Manila the previous week: the Allies would submit 'requirements' through Japanese government 'channels.' There would be no need to kick down doors or conduct harsh interrogations for answers. In any event, Willoughby warned: 'The enormous initial discrepancy between US occupation forces and available Japanese major units is so complete that punitive or disciplinary features are impracticable now and may become fatal, if initiated prematurely.'[16]

Fortunately, for both the Allied forces and the Japanese, the Manila understanding held. Cooperation, much of it conducted between American and Japanese intelligence officers, displaced coercion and punishment in Occupied Japan. The British Commonwealth Occupation Force (BCOF), responsible for the occupation of Shikoku and southern Honshu, was fortunate to operate in a cooperative environment. *BCON*, the BCOF newspaper, reported in August 1946 that Commonwealth investigations had 'found huge arsenals of arms and ammunition in vast caves and underground factories throughout the area.' In Hiroshima Prefecture, BCOF units uncovered 100,000 tons of explosives and stores of poison gas at a production plant. US forces, too, were finding 'thousands of caches located deep in the hills, in carefully constructed tunnels and warehouses.' Indeed, according to a history of those years, 'every month of the Occupation disclosed new caches of military supplies.' Commandos of the Nakano School, tasked with organizing guerrilla forces for Japan's planned decisive battle at the end of the recent war, would have led irregular groups armed with cached weapons against a punitive occupation.[17]

The US Army immediately sent scientists and technical experts[18] to Tokyo and dispatched CIC teams around Japan to investigate military and naval technical research institutes, equipment, projects, and personnel. Compton's team reported on 1 November their findings, as members of the United States Army Forces, Pacific (AFPAC) Scientific and Technical Advisory Section, their 'Report on Scientific Intelligence Survey in Japan.' Excluding the fields of aeronautics, military medicine, and atomic energy, left for study by 'other agencies', the Compton group had acted in the first two months of the occupation to investigate military and civilian R&D organisations, assess their effectiveness, and 'ferret out... any new discoveries or techniques of significance' for further study.

The report uncovered some information on Noborito. The survey team learnt that the 9th ATRI was 'primarily concerned with special equipment for subversive agents, special police.' The investigators reported that Noborito had developed small radio transmitters, 'characterised by simplicity in construction and operation, lightness of weight, and small size.' They also noted Noborito's work on the death ray and the bombing balloons. Investigators had queried key officials, including Major General Kusaba Sueki, the director of Noborito's project to develop the balloons.[19]

In addition to interviewing key Noborito officers, the US Army sent teams within weeks of the Japanese surrender to Noborito and the institute's branches in Nagano and Hyogo prefectures. Japanese witnesses to their arrival recalled that the investigators were from the CIC and identified some of the men as Nisei (Japanese Americans born to immigrant parents).[20]

Captain Takiwaki Shigenobu, a technical officer in Second Section, witnessed CIC officers arrive at Noborito in September and leave with materials and documents. At Matsukawa National School, which served late in the war as the headquarters for Noborito's Hokuan Branch, the entry of the school's journal for 5 October noted the arrival of two

American soldiers to inspect branch facilities. On 1 November, according to the journal, soldiers returned to confiscate equipment and materials.[21]

In the village of Ina, Nagano Prefecture, Second Section's Major Ban encountered in late October two Nisei officers from the 441st CIC Detachment, sent to investigate the site after Lieutenant General Shinoda had submitted a report mentioning it. The meeting was a cordial one. Ban felt no pressure as the CIC officers questioned him. The Nisei officers confiscated various items and ordered him to destroy explosives discovered there. Once Ban had done so, the visitors shook his hand, thanked him for his cooperation, and left.[22]

Handling Allied Demands for Intelligence

Lieutenant Colonel Niizuma's order of 15 August, issued more than two weeks before Japan's formal surrender and occupation, to destroy all incriminating evidence of special research had left behind only documents and materials of secondary importance. For detailed intelligence, investigators had to submit requests through Japanese channels to speak to those with the answers to their questions.

GHQ/SCAP established early in the Occupation a procedure for gathering Japanese technical intelligence. The Technical Liaison and Investigation Department (TLID), which operated under the Office of the Chief Signal Officer (OCSigO), or some other unit would submit a written request to the War Department Intelligence Target Section (WDIT) of G-2 to interview persons of interest. The request would then go to Japanese officials to process.

In one case, WDIT received on 10 December a request that Noborito's Lieutenant General Shinoda, Major General Kusaba, and Colonel Yamada appear before officers of the 5250th Technical Intelligence Company (TIC) of the US Army Technical Intelligence Center at the former Tokyo First Army Arsenal for a half day of questioning on 'powerful ultra-shortwave' research and other Noborito projects. On 11 December, Colonel Teshima Haruo, who had served in Chile as military attache to the Japanese embassy and headed the Nakano School's testing unit before working under Arisue after the surrender, sent the request, translated into Japanese, to Major Shishikura Juro, who had worked in intelligence in the Kwantung Army and was serving as another of Arisue's men, with instructions to assign the case to Lieutenant Colonel Kusakari Michitomo, who had served at the

AOAH and had relayed Niizuma's order of 15 August to Noborito to destroy all incriminating evidence.

The document of 10 December noted that the interviews were to 'crystallise further plans for investigating this Institute.' Subsequent conferences included a discussion on 29 December with Major General Kusaba on the bombing balloons and a three-day visit in January 1946 to Noborito's Ina and Nakazawa branches, as well as the affiliated Japan High-Frequency Company, to discuss details of the death ray with the company's Dr Ikebe Tsuneto and have a talk with Noborito's Major Ban on 'photographic projects and developments.'[23]

The Americans, who had avoided a possible bloodbath in Occupied Japan by opting for cooperation over punitive coercion, found the quest for intelligence slow going as Arisue and other IJA officers brokered Japanese information. Arisue and other officers met Japanese summoned for interviews and guided them on what to say. In short, the brokers worked to keep incriminating evidence hidden but cooperated when revealing secrets would result in no disadvantage. The general advice, or common sense among those questioned, appears to have been to deny knowledge of a sensitive issue until presented with irrefutable proof.

On 10 June 1946, Major Ban sat for questioning at the Tokyo headquarters of the 5250th TIC. He was there to discuss secret inks following a striking intelligence discovery. In January that year, the Civil Censorship Detachment (CCD) of GHQ/SCAP's Civil Intelligence Section had obtained copies of three 'espionage manuals' that, according to a 21 February report of Colonel Harry I.T. Creswell, CCD's counterintelligence chief, constituted the 'first evidence that the Japanese have knowledge of and use secret writing.' Creswell's report included a research paper of CCD's Special Activities Division (SAD), according to which American intelligence had until then 'no conclusive proof that the Japanese had ever used Secret Inks (S/I), or other Secret Writing (S/W) methods for espionage purposes.'

CCD had intercepted on 16 January that year a package of 17 classified IJA documents, originating with 12th Section (War Guidance) of the Army General Staff in Tokyo and addressed to 2nd Section (Intelligence) of IJA Chosen Army Headquarters. Three of the documents concerned methods of secret writing; Noborito had conducted the research. SAD concluded on the basis of the three documents that this 'first evidence' of secret writing proved 'beyond a reasonable doubt' that the Japanese Army had 'spent considerable time and research effort to develop suitable and good methods.'[24]

Ban admitted in the course of his conversation that June day with CCD and TIC officers to not having 'given a detailed account' in a previous interview of his work, which had included secret inks, to Lieutenant Henning, sitting before him again. The TIC officer had questioned him the previous year at the Noborito branch in Ina Village, Nagano Prefecture. The reason, Ban explained, was that 'it would have proved detrimental to the interests of both me and my country.' Ban further admitted that he had consulted with his wartime superior, Colonel Yamada, before sitting 'on this occasion' for questioning.

The interview of 10 June, in which Ban admitted to secrets only when presented evidence of them, was typical of the intelligence challenge facing GHQ/SCAP. From the record, it appears that Ban continued that day to withhold intelligence. He said nothing of IJA biological warfare, another area in which had worked. He addressed only aspects of secret writing put to him in the meeting. In June 1946, as far as Ban knew, volunteering information on biological warfare would have been detrimental to himself and to Japan.

One of Ban's admissions highlighted the eagerness of US intelligence officers to obtain Japanese intelligence. In the course of the meeting, Ban repeatedly sang the praises of his subordinate, Captain Arikawa Shunichi, as a master of secret inks. His praise led a TIC officer to go fetch Arikawa for a separate meeting. The Americans at first thought little of Ban's subordinate, who was 'shabby' in appearance. In the course of the interview, according to the subsequent report, Arikawa 'proved himself to be brilliant.' Impressed, the Americans offered him lunch and cigarettes. Having brought him in a jeep for questioning, they sent him home in a passenger car. Arikawa, for his part, felt no pressure to keep his work secret, having worked on secret inks rather than the pathogens and poisons that others in Second Section had developed. What he had to say was not detrimental.[25]

Niizuma in the Loop

Lieutenant Colonel Niizuma was often in the loop when the Americans were seeking technical intelligence. After ordering the destruction of all incriminating evidence of special military research, above all that related to the Fu-go project for bombing balloons and other projects at Noborito, he had joined Japan's Central Liaison Office (CLO), the key office for handling requests from GHQ/SCAP.[26] His position allowed him to appear at various interviews that members of Dr Compton's survey team and others conducted in the early weeks of the Occupation. One can imagine that he worked to ensure that IJA officers meeting Americans for questioning stuck to a script prepared in advance.

Niizuma apparently worked to deceive the Americans and deny them intelligence on sensitive topics. On 13 October, he accompanied Dr Arakawa Hidetoshi of Japan's Central Meteorological Observatory to his interview with members of Compton's team. Arakawa had been a key player in the Fu-go project. Kusaba cited Arakawa by name in a postwar article for his 'outstanding efforts' in the fundamental work of charting the North Pacific jet stream that the paper weapons rode from Japan to the United States. Arakawa, with Niizuma at his side, proved in the face of many detailed questions 'unable to furnish any exact information on his studies in this field.'

Niizuma also accompanied Kusaba to a meeting in late November to answer questions on the bombing balloons. In the early months of the occupation, Niizuma accompanied other officers or went alone to meetings on the topics of Noborito and IJA programmes to develop biological and chemical weapons.

In late October, Niizuma had answered a summons to discuss Noborito with Edward L. Moreland, who had replaced Compton as chief of the Scientific Intelligence Survey. The following month, another scientist at GHQ called on Niizuma to discuss the Fu-go project. Apart from Noborito issues, Niizuma in the Occupation's early months attended meetings to offer answers or accompany other Japanese military officers summoned to interviews on IJA biological and chemical warfare projects and actions.[27]

Sticking to the Script

IJA officers outside of Noborito also seemed to recite from a script. They denied the Americans intelligence on the biological weapons developed for the bombing balloons and claimed the project had existed only for propaganda. On 19 September, Lieutenant Colonel Kunitake Teruhito from the Army General Staff and Major Inoue Shigeru, commanding officer of the balloon regiment that launched Noborito's bombing balloons, in speaking to CWS officers Lieutenant Colonel Murray Sanders and Major Howard Skipper, offered some basic information on Noborito's bombing balloons but insisted, in the language of report, that 'it had never been the Japanese intention to alter these loads with CW or other weapons.' They admitted that the Japanese Army had launched approximately 9,000 balloons over a period of six months, but spoke of the programme as a mere exercise to lift Japanese morale after the 1942 Doolittle Raid. They further claimed that the programme would not have continued if the war had lasted longer. Kunitake and Inoue said nothing concerning the rinderpest and other pathogens prepared as payloads or the balloon production that continued until the day of the emperor's broadcast ending the war.[28]

Other IJA officers repeated the same points: the Japanese Army ordered the Fu-go project as a propaganda exercise to regain 'face' following the humiliating Doolittle Raid of 1942; the conventional explosives would have no major military effect but would spread fear in the United States; the IJA cancelled the programme in the spring of 1945 because American censorship had denied them their propaganda victory.

Kusaba met in October with members of the survey team and submitted a report on the bombing balloons. Kusaba maintained the line given earlier by Kunitake and Inoue, claiming that the Japanese Army had ceased conducting further research on the programme in April and did not intend to launch further attacks.[29]

Three years later, Kusaba repeated his story in 1948 in a meeting in Japan with Brigadier General William H. Wilbur, who had directed defences against the balloons as chief of staff of the Western Defense Command. Wilbur, apparently none the wiser, included Kusaba's deception in an article, 'Those Japanese Balloons,' published in August 1950 in the popular monthly magazine *Reader's Digest*. Goto Masao, who had served in the war as a councillor on the government's Board of Technology, informed Japanese readers in a 1956 special edition of the popular weekly *Shukan Yomiuri* that Wilbur had committed a 'great error' in accepting Kusaba's story. In fact, Goto revealed, the Japanese Army had only ceased releasing bombing balloons in April 1945 due to the seasonal lull in the jet stream, which 'ceases entirely in summer.' Meanwhile, in anticipation of a resumption of the bombing campaign in the autumn, balloon output had continued until the surrender.[30]

Assessing the Death Ray

Early in the occupation, Japanese intelligence brokers appeared forthcoming in providing technical intelligence on projects little likely to result in criminal trial for those involved or to damage Japanese interests. This seems to be the case with the death ray, details of which soon emerged. After all, Noborito had never completed the project. Only lab animals died in the course of its experiments. It was safe to reveal details of the institute's Ku-go project.

Given relatively abundant intelligence on Noborito's death ray early in the occupation, US army and navy officers produced within the first months of the occupation a number of reports on the project. Lieutenant General Shinoda and Major General Kusaba sat for questions. Drs Sasada Sukesaburo, who had directed the animal experiments, and Ikebe Tsuneto, an engineer at the Japan High-Frequency Company, were among the Noborito technicians and contractors who answered questions and submitted written reports. Investigators paid at least two visits in the first half of 1946 to Japan High-Frequency's facilities in Yokohama and Nagano Prefecture. The Japanese interviewed at the company's two sites produced technical data and equipment for the investigators. 'Magnetron and other tubes' found 'to have technical intelligence value', according to one report, went to the Intelligence and Communications Coordination Branch at the Holabird Signal Depot in Baltimore, Maryland, for analysis. Osaka tubes, developed for Noborito by a contractor at Osaka Imperial University, went to the Naval Research Laboratory.[31]

The investigators were impressed with what the Japanese showed them. Those who had interviewed Shinoda and Kusaba saw potential: 'With the development of higher-power and shorter-wave length oscillators, which has become possible through the Allied research on radar, it is possible that a death ray might be developed that could kill unshielded human beings at a distance of five to ten miles if these Japanese experiments are reliable indications of the potentialities of the death ray.'[32]

From Pusan to Pyongyang

Following Japan's surrender, the United States Army Military Government in Korea (USAMGIK) had replaced Japan's Office of the Governor-General of Chosen (OGGC) in southern Korea, the American zone of occupation south of the 38th parallel. The US military took control in Fusan (known after Korea's liberation as Pusan) of the OGGC Livestock Hygiene Research Institute. In November 1945, the authorities appointed Dr Nakamura Junji, who had been directing the institute's Virology Department, technical advisor to the institute. He and Dr Ochi Yuichi, who had become the institute's director near the end of the war, were denied permission to return to Japan until the institute's new Korean managers had learnt the ropes.[33]

Nakamura, eager to return to Japan and establish his own research institute, disobeyed the American order. He sent in secret two of his subordinates – Iwasa Takeshi and Miyamoto Takeshi – by smuggling ship back to Japan. In Tokyo, they presented Nakamura's proposal for a new institute to Dr Kasai Katsuya, who had consulted for the OGGC's institute and was directing a department of the influential Kitasato Institute in Tokyo, and to Dr Nakamura Norichika, director of the MAF Animal Infectious Disease Research Laboratory.

The two subordinates informed Nakamura on their return to Pusan that his proposal had received favourable responses. Nakamura then gained the tacit approval of the institute's new Korean director, Dr Kim Jong Hui, his 'favourite apprentice' during the war, to make his own discrete trip to Japan on the pretext of going home to celebrate the New Year. In Tokyo, Nakamura brought Iwasa and Noborito's Major Kuba Noboru, to a meeting on 7 January 1946 to receive approval for his planned institute in meetings with directors of the Japan Veterinary Medical Association. Kasai christened Nakamura's future organisation the Laboratory of Comparative Pathology (LCP).[34]

In Korea, Kim Jong Hui was furious. Handed the reins to the institute and running it with the advice of former superiors Nakamura and Ochi, Dr Kim soon ran into interference from USAMGIK. Kim wrote years later that 'American invaders, bringing gangsters with them,' took away technical documents, installed a US army captain over him as his 'advisor,' and 'plotted' to take Kim and other institute personnel to the United States. In February 1946, a 'friend' from the Livestock Department of the Agriculture and Forestry Bureau in the North Korean Provisional People's Committee (NKPPC) in Pyongyang came south to see Kim on orders of Kim Il Sung, requesting that he provide serum from the institute, the only one of its kind in Korea, to treat cattle plague in Soviet-occupied northern Korea. Kim sent north a 'large quantity' of the serum. US authorities, on learning what he had done, interrogated him, detained him for several months, and expelled him from the institute. In November that year, indignant and unemployed, Kim packed some 'bacterial strains and biologics' in a trunk and headed north with his wife and children to Pyongyang.[35]

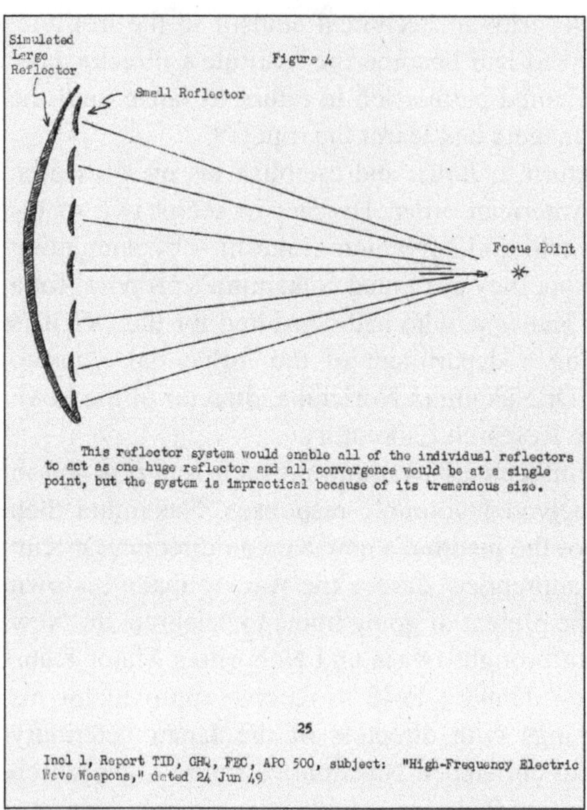

Left: Illustration from a 1949 US military intelligence report, *Japanese Research on High-Frequency Electric Wave Weapons*, showing part of a death ray that Noborito was developing at the end of the Second World War. (Courtesy of the US National Archives)

Below: Postwar map of the northern part of Noborito. The map was attached to a request of the Kawasaki Repatriates' Welfare Association, dated 25 November 1946, to use two of the site's buildings to house Japanese repatriated from abroad. (Courtesy of the US National Archives)

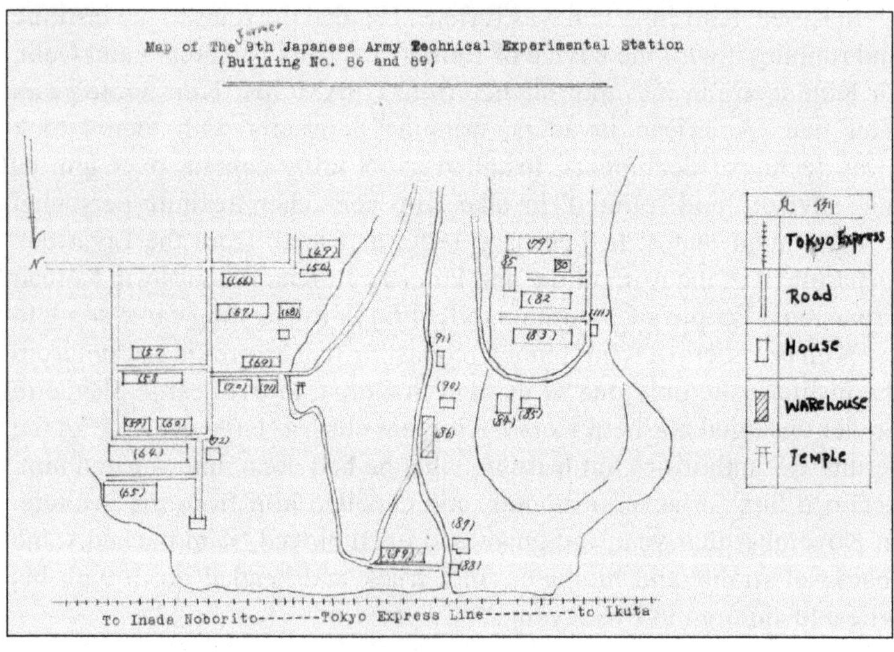

CHAPTER 8

WORKING WITH THE AMERICANS

> The worksite was on the US naval base at Yokosuka. The pay was good, so I agreed to join. Along with that, when all is said and done, what most impressed me was coming into contact with and getting to know the Americans. They didn't look down on us technicians as people from a defeated country but treated us as partners.
>
> – Major Okada Masayuki, former chief of Centre Group, Third Section, Noborito Research Institute[1]

The victorious Allied forces occupied Japan with a light touch. GHQ/SCAP officials appropriated a number of private and public properties for the occupation forces; jailed a few top Japanese civilian officials and military officers for trial as Class A war criminals; gathered military technology and hardware for research and exploitation; abolished the Japanese Army and Navy, including their intelligence organisations; disbanded the Home Ministry, including its powerful police; and did away with the intelligence organs of the Ministry of Foreign Affairs, including its consular police and the Radio Room for monitoring foreign broadcasts. From the start, however, Washington approached Tokyo less as a vanquished foe to punish than as a junior partner against the Communist Bloc.

Moscow occupied no territory in the Japanese home islands and played no important role in administering Occupied Japan. Even before the formal surrender, Japanese officials could see that Washington and London's wartime alliance with Moscow had effectively ended when the fighting stopped. At Atsugi air field on 28 August, Lieutenant General Arisue Seizo had greeted General Douglas MacArthur's advance party to discuss details of the occupation. Pointing to two 'loitering' Soviet military attaches from the embassy, Arisue asked if the Soviet officers were with the Americans. A US officer dismissed the idea in replying that the Soviets had nothing to do with them. At that moment, Arisue recalled in one of his memoirs he had a 'hint' of how the postwar period would play out. In a strategic environment of opposing blocs, the Japanese would pursue their interests by partnering with the Americans.[2]

Growing tensions in the Far East brought GHQ/SCAP together with various Japanese intelligence operators. The postwar division of Korea into US and Soviet zones of occupation hardened in 1948, with Seoul claiming control of the entire peninsula and establishing on 15 August of the Republic of Korea (ROK). Pyongyang in turn proclaimed the following month rule over all Korea in founding the Democratic People's Republic of Korea (DPRK). The stage was set for civil war. The Chinese Nationalist loss of Manchuria in 1948 to the Chinese Communists was a major blow that foreshadowed the Nationalist retreat to Taiwan and the Communist victory in Mainland China the following year.

In Tokyo, the Americans turned to Japanese veterans for Asian intelligence, particularly for intelligence on the Korean Peninsula, Manchuria, China, and the Soviet Far East. With the intelligence organs of Imperial Japan disbanded under the Occupation, many former intelligence officers and operatives went to work for the Americans. Pay was one motive. Another was the calculation that working with the United States to preserve Japan's Imperial Household and later revive the imperial army and navy.

The Americans, for their part, appreciated Japanese expertise as developments in Asia increasingly turned in favour of the Communist Bloc. They recruited intelligence talent in Japan as the Communists prevailed over the Nationalists in China's civil war, armed clashes erupted along the inter-Korean border, and the powerful Soviet Union loomed as a potential adversary in a global war that could come at any time. For communications intelligence (COMINT), a key intelligence source, the US Army recruited IJA veterans to work as Russian linguists and code breakers at a number of sites in Japan. Pressed for linguists in the Korean War, according to a history of the US National Security Agency (NSA), the US Air Force Security Service brokered a deal with a General Hirota, former chief of IJA COMINT, for 'a dozen Japanese linguists who were fluent in Chinese.'[3]

The Americans also backed IJA veterans in secret intelligence operations in Japan and across continental Asia, from Korea in the east to Pakistan in the west. Major General Charles Willoughby oversaw a large intelligence programme directed against foreign targets (code name 'Take,'

Japanese for 'bamboo') and Japanese communists and other leftists (code name 'Matsu,' meaning 'pine') under the collective code name 'Takematsu.' Japanese intelligence veterans in some cases established trading corporations and sent abroad operatives who smuggled goods in and out of Occupied Japan as cover for intelligence activities.

Many of the Japanese recruited into American operations were officers with combat and command experience, but some were senior intelligence veterans. Among them were Lieutenant Generals Arisue and Kawabe, who had met Willoughby at the start of the occupation.[4]

Recruiting Document Specialists

Lieutenant General Arisue played the role of intelligence broker one day in the spring of 1948. Colonel Yamamoto Kenzo, who had directed Noborito's counterfeiting section, called on him for advice. He had received a summons to appear for questioning at the NYK Building, which housed offices of GHQ/SCAP's G-2. Yamamoto, unsure what he should say, sought Arisue's guidance. Should he mention Noborito's wartime production of counterfeit Chinese bank notes in Operation Sugi? If so, how much should he reveal? Arisue counseled Noborito's chief counterfeiter, 'Tell them everything.'

So advised, Yamamoto took a streetcar to a stop near Tokyo Station, stepped into the street, walked into the nearby NYK Building, and proceeded to the fourth floor for an interview with American intelligence officers. A young Nisei officer opened the session, then introduced a more senior field officer to continue the meeting.

Yamamoto surprised the officers by telling them details of Third Section's counterfeiting. He claimed in his memoir that the astonished interrogators could scarcely conceive that he, an intendance officer, had directed such sensitive production. Intendance officers typically handled everyday pay and supply issues. Yet he had directed some of Noborito's most secret work.

Yamamoto had received his summons long after the Americans had interrogated Noborito's other senior officers in the first months of the Occupation. Had the Japanese War succeeded in hiding Yamamoto's activities until then? Possibly. The War Ministry's AOAH had produced a document, dated 31 August 1945, on Noborito but referred only to the bombing balloons and some other activities of First Section. Missing were any references to the BW projects of Second Section and the counterfeiting of Third Section.[5] The Japanese Army had kept from the Americans for some time intelligence on Noborito's counterfeiting. How much did the Americans know before talking to Yamamoto that day? The Chinese had discovered in the war that Japan was circulating bogus Chinese bank notes. The British, having captured Indian agents sent into India, almost certainly with counterfeit rupees in their possession, must have detected them. Perhaps, while knowing of Japanese wartime counterfeiting, the US Army had been unaware for a time of Noborito's role in general and that of Colonel Yamamoto in particular.[6]

Yamamoto revealed a great many details over the course of a month's questioning. US officials, impressed from the first, invited him to cooperate by sharing what he knew about counterfeiting. The Americans also wished to learn the details of what he knew about topographical intelligence on the Soviet Far East along the Manchurian border, where he had once conducted surveys. In discussions, the Americans spoke of 'give and take.' In addition to paying him to work for them, they offered in return to protect him from any Soviet or Chinese demands to prosecute him for war crimes. Yamamoto at first made no commitment but in the end agreed to cooperate. As was the case with officers of Unit 731 and other IJA organisations involved in biological warfare, all were spared prosecution in Occupied Japan, Yamamoto never went on trial for his counterfeiting. In the words of Igarashi Nobuo, a technician who worked under Yamamoto in the Second World War and with the Americans in the Cold War, Washington was eager to obtain Japanese technology and ready to grant immunity from prosecution in exchange: 'None of them were made war criminals because America wanted the technology and sought to use it themselves… With someone like Colonel Yamamoto Kenzo, it would not have been odd for him to have been made a war criminal.'[7]

Scouting counterfeiting talent was evidently part of the agreement. In the spring of 1950, several months after the Chinese communists had established the People's Republic of China and as the Korean communists in Pyongyang were preparing to invade the Republic of Korea, Yamamoto called on former subordinates from Noborito's Third Section to join him in working for the United States. In May, the month before the Korean People's Army (KPA) advanced south to re-unify the peninsula, Yamamoto and 10 or so of his former subordinates started working inside Yokosuka, the former IJN naval base located near the entrance to Tokyo Bay. One of Yamamoto's people was Iwase Kiyoshi, to whom he had sent earlier that month a letter proposing employment. Yamamoto had offered little in the way of details but had promised a high salary. It was enough to convince the young man to accept his old boss's offer. In June, the month that the Korean Peninsula erupted in war, Yamamoto recruited Okada Masayuki. Visiting his former subordinate at his home in Osaka, Yamamoto had swayed him by offering generous payment for his skill at analysing paper and other materials.

The sprawling naval facility was an excellent site to hide Japanese engaged in secret work. The Noborito veterans would enter the base through a gate used by many Japanese with regular jobs at Yokosuka, then proceed to an unremarkable building containing an organisation with the blandest of names to hide its secret purpose: the Government Printing Supplies Office (GPSO). The building, two stories high, included rooms for analysis, paper making, and printing. American officers and technicians, some of them Nisei, worked there with Yamamoto's men. As one Noborito veteran who worked at the GPSO explained, putting a putative supply office inside a large naval base gave the outfit excellent cover. The people and supplies going in and out of the building looked like what one would expect to see in a normal logistics or supply centre. Inside the building, Noborito veterans forged passports and other documents.

Colonel Yamamoto's men came to the GPSO with solid backgrounds. Okada Masayuki, an IJA technical major, had worked under Yamamoto as chief of Third Section's Centre Group, which had handled analysis, forensics, and inks. Okada had graduated in 1940 with a degree in applied chemistry from the Faculty of Engineering at Japan's prestigious Osaka Imperial University and joined Noborito on the advice of his professor. In the GPSO, Okada analysed paper, fibre, and ink. Dedicated to his craft, Okada would visit fabric wholesale shops around Yokosuka on weekends and purchase materials to use at work, such as silk cloth for fabricating foreign military notebooks.

Working with Okada was Iwase, one of his former subordinates in Centre Group. Forging KPA military notebooks was one of Iwase's tasks at the GPSO. He would take apart authentic notebooks sent from Korea, analysing such elements as the quality of the paper, the composition of the ink, and the style of printing before making copies of them. Iwase found such work easier than what he had done at Noborito.[8]

From the GPSO in Yokosuka, forged documents went into storage in facilities located in Japan and Korea. Together with captured or reproduced uniforms and weapons, Korean military operatives took GPSO documents with them on missions behind the lines in Korea and China. The appropriate documents helped to give each operative an authentic appearance, completing such disguises as an enemy soldier, an internal security officer or perhaps a labourer. When challenged to produce an internal travel document or military ID, an operative's life depended on the quality of GPSO production. The Korea Liaison Office (KLO), General Willoughby's organisation of Koreans operating in enemy territory, apparently took GPSO documents on their missions. A Japanese researcher on Noborito's history wrote, without further detail, that the Canon Agency, an organisation for operations in Japan and abroad, led by Colonel Jack Canon in Tokyo under General Willoughby's G-2; the 308th CIC Detachment of the US Eighth Army; and Korean partisan forces operating under the Combined Command for Reconnaissance Activities, Korea (CCRAK) and Combined Command for Reconnaissance Activities Far East (CCRAFE) received fabricated materials from the GPSO.[9]

The ROK military operatives executing intelligence missions behind enemy lines on DPRK defences for several weeks before MacArthur's bold amphibious assault of 15 September at Inchon likely carried GPSO forgeries in the pockets. Disguised as KPA soldiers, fishermen, dock workers, and such as they gathered intelligence,[10] they needed military notebooks, identification cards, travel documents, and other bits of paper known collectively in the world of espionage as 'pocket litter' to survive. KPA uniforms were widely available, but sophisticated forgeries for surviving in enemy territory required experienced counterfeiters working with the proper paper, ink, and equipment. Veterans of Noborito's Third Section in the GPSO likely made the documents for the operation.[10]

Following Allied success at Inchon, Lieutenant Eugene Clark, a naval officer who had been compiling hydrographic intelligence in GHQ/SCAP G-2's Cartographic Branch in Tokyo, set sail in October for the mouth of the Yalu River in advance of General MacArthur's northern advance to the Chinese border. With him were Colonel Kye In-Ju of the ROK Army and Lieutenant Yon Chong of the ROK Navy. The two men, attached to G-2 in Tokyo, both spoke excellent Japanese. Yon spoke good English as well. The two men had joined Lieutenant Clark, who spoke Japanese, the previous month on a reconnaissance mission that took them to the island of Palmido outside Inchon prior to the invasion. In October, they led approximately 150 Chinese and Korean operatives to gather intelligence on Chinese and Korean communist troop movements in areas near the Sino-Korean border. Yon recalled landing some operatives to go gather intelligence around Dalian and the naval port of Lushun at the tip of the Liaodong peninsula in northeast China. Others headed for Sinuiju, the Korean city on the Yalu River across from Tantung, the main gateway between the Korean Peninsula and northeast China.[11]

For such dangerous work behind enemy lines, the operatives carried counterfeit documents. Yon handed out documents to go with the disguises – uniforms of the PLA or KPA, as well as clothing appropriate for a Chinese coolie or a Korean peasant – and weapons. Some of the identification documents even came with photographs of the operatives, according to Yon, which suggests document forgeries rather than the simple distribution of genuine documents collected from prisoners and the dead. It had all come from Tokyo. Yon recalled that G-2 in Tokyo supplied the documents, clothing, and weapons. A Japanese intelligence historian wrote that Yon in his postwar account of the operations was 'alluding to' the work of Noborito's Third Section veterans at the GPSO.[12]

Korean operatives also relied on Tokyo to equip them for missions to infiltrate North Korean hospitals to gather intelligence to confirm or dispel rumors that the plague was spreading in the north. ROK Army Colonel Choe Kyu-bong recalled that the KLO received prior to a mission in 1951 to infiltrate a hospital in South Hamgyong Province, 'as in the past,' identification documents made in Tokyo. KLO, having sent samples of recently obtained identification documents of the KPA, Workers' Party of Korea, and Ministry of National Defence, had requested that 'Tokyo' reproduce them and insert photographs of its operatives in them.[13]

Lieutenant Yon, too, needed forged documents in his mission to search for signs of plague in another medical facility, a DPRK field hospital in Wonsan. In March 1951, Yon joined Lieutenant Clark and Brigadier General Crawford Sams of the US Army Medical Corps on the mission. While Clark and Sams hid near the landing site, Yon, disguised as a KPA first lieutenant, led two junior officers and 60 men in disguise to the hospital. Members of his group subdued the dozen guards around the hospital and took their places. Yon, joined by an operative posing as a nurse, entered the hospital and a separate tent for corpses, selected three patients and two corpses for Sams to examine, commandeered a jeep and an ambulance to transport them, then headed back to the hiding place. On the way, Yon's group ran into a KPA security detachment that outnumbered them. Thanks to the disguises, Yon was able to bluff his group past the enemy and deliver the patients and corpses to Sams. The mission was a success. Sams was able to diagnose the supposed plague as another disease, a less alarming one.[14]

Working With Uncle Sam

As Colonel Yamamoto had promised his men, the working conditions in Yokosuka were excellent. American personnel from the GPSO and other units treated the Noborito veterans with respect. Americans addressed Yamamoto, the senior Japanese member of the GPSO, as 'chief' and 'colonel.' Major Okada found that the Americans treated them not as inferiors from a defeated enemy nation but as colleagues. The Noborito veterans enjoyed high salaries at a time when most Japanese were making do with little in a nation still recovering from defeat in war. Working on a US naval base with Americans, Yamamoto and his men enjoyed America's postwar material abundance, including access to food, drink, and cartons of Camel and Lucky Strike cigarettes that were largely unavailable outside the base.[15]

The Noborito men who were working with the Americans at the GPSO were far from the only Japanese who contributed to US military operations in Korea. Some of the thousands of Nisei who had served on the Japanese side against the country of their birth in the Second World War worked after Japan's surrender for the United States in Occupied Japan and Korea. They served Uncle Sam in Japan as censors, interpreters, and translators. In the Korean War, they interpreted and interrogated prisoners.[16]

Japanese surveyors with backgrounds in the IJA's Imperial Land Survey (ILS) and other organisations worked under the US Army's 64th Engineer Battalion (Base Topographic), which produced military maps of Korea, Manchuria, and other sites on the third floor of the Isetan Department Store in Tokyo's Shinjuku District before moving to a new site near Jujo Station. With the 64th Engineer Battalion, they combined data from the ILS with postwar aerial photographs of the US military to produce topographic maps for military campaigns of the Korean War.[17] Thousands of Japanese worked in Occupied Japan for US intelligence organisations, including the more than 5,000 Japanese censoring the mail of their fellow countrymen in 1949 alone for SCAP's CCD.[18]

In another case, Prime Minister Yoshida Shigeru ordered in October 1950 a reluctant Okubo Takeo, chief of the Japanese Maritime Safety Agency (MSA), to send ships to clear mines from Korean waters for the US Navy to carry out the amphibious invasion of Wonsan and other operations. As Okazaki Katsuo, the prime minister's chief cabinet secretary, explained to Okubo, Yoshida had decided to send MSA ships to sweep Korean waters for mines as a way to gain advantage for Tokyo in negotiating Japan's future peace treaty.[19]

Shuttling Between Yokosuka and California

In June 1952, almost a year after representatives of the UN Command began with counterparts of the KPA and the Chinese People's Volunteers (CPV) the truce talks that would end in the 1953 Korean Armistice Agreement, Colonel Yamamoto led several Noborito veterans on board a US military aircraft at Haneda Airport for a flight to San Francisco. All had volunteered to continue their work in the United States. Among them were Okada and Iwase. A second group left Japan in August, followed later by a third. In all, a dozen Noborito veterans followed Yamamoto to the United States to continue working at a new base of operations: a naval facility in San Bruno, south of San Francisco in California. As at Yokosuka, the Japanese document specialists worked at what appears to be an unassuming office of no particular interest. One of the men remembered it as a 'naval supplies office.'[20]

Succeeding Colonel Yamamoto as chief of the Japanese unit at Yokosuka that year was Major Ban Shigeo. The former chief of Group 1 in Noborito's Second Section, Ban in 1950 was working for Taimei Chemicals Co., Ltd., a company located in Nagano Prefecture. The site was near where some members of the Noborito Research Institute had withdrawn late in the war to prepare for Japan's last stand. A member of the board of directors, Ban held concurrent positions as chief of the company's laboratories and director of its technical department. In 1951, working on a 10-year contract, Ban became chief of the GPSO's chemical section. Following Yamamoto's departure for California the next year, Ban took his place as the senior Japanese in the GPSO.

At Yokosuka, Ban recruited more Noborito veterans to join the GPSO. In addition to former members of Yamamoto's Third Section,

Ban recruited from his own Second Section, hiring veterans skilled in techniques of secret writing and photography. One of them, Maruyama Masao, a technical major, had headed Second Section's Group 5, which was responsible for developing spy cameras, microdots, and other technology. Some of the new hires had no Noborito connection. In all, Ban brought in about 20 people. In hiring veterans from Second Section in addition to the counterfeiters of Third Section, Ban expanded the scope of the Japanese contribution to American operations. Where the counterfeiters worked on agent authentication by fabricating documents, veterans of Second Section would have contributed to agent operations with their work in such fields as microdots, miniature cameras, and secret inks.

In 1955, Ban flew to San Francisco to provide technical guidance, presumably for a two-year tour. During that decade, GPSO members shuttled back and forth between Yokosuka and San Bruno.[21]

Closing a Base, Transferring Personnel

From 1952 to 1961, Japanese members of the GPSO shuttled back and forth between Yokosuka and California on two-year assignments. When Chinese forces pushing south in late 1950 seemed capable of overrunning the entire Korean Peninsula, there was talk of moving the GPSO to the United States. According to the technician Igarashi Nobuo, concern over Yokosuka's possible vulnerability to attack led to thoughts of 'moving the work to America.' The return of full sovereignty to Japan in April 1952 under the terms of the 1951 Treaty of San Francisco was possibly another factor for sending GPSO Japanese members to California on temporary duty.

In 1960, the decision was made to close the GPSO in Yokosuka. The advent of the 1960 Treaty of Mutual Cooperation and Security between Japan and the United States of America, with its status of forces agreement (SOFA) governing US use of Japanese facilities, possibly was a factor in the decision. In 1961, the GPSO closed its doors. Many of the Japanese retired at that time. Ban, who had reached the end of his contract, became an executive director at Taimei Chemicals. Other GPSO members continued working in California.[22]

In California, as some of the personnel came from Noborito, so did some of the equipment. Iwakuro Hideo, who played a key role in establishing Noborito, made Yamamoto chief of Third Section, and wielded influence in Occupied Japan as a broker of intelligence, told an interviewer in 1967 how IJA officers early in the occupation had taken apart and hidden a Sammel printing press used in Noborito's Third Section. Americans discovered the cached parts but did not recognise them for what they were. Noborito veterans later put the press back together to use in partnership with the Americans. The equipment probably ran once again at the GPSO in Yokosuka, then in California. As Iwakuro told the interviewer with a laugh, years after the GPSO had closed its doors, 'It now seems that it has once again been assembled and put to use!'[23]

At the naval facility in San Bruno, Iwase recalled, the work was 'an extension' of what they had been doing at Yokosuka. The Japanese technicians worked on 'many kinds' of reproductions, among them passports and military documents of countries in Southeast Asia. 'Particularly numerous,' according to Iwase, were 'Communist Chinese (PRC) documents.'[24]

The 1953 Armistice, which ended the battlefield fighting in Korea, came as conflict in Indochina grew heated. The Japanese counterfeiters turned their attention increasingly to Southeast Asia as Washington's involvement in the region grew after France's 1954 defeat at Dien Bien Phu and Washington's backing the following year the creation of the Republic of Vietnam in Saigon to counter the Democratic Republic of Vietnam (DRV) in Hanoi. One of the Japanese technicians, a former member of Noborito's Third Section recruited in 1953 by Yamamoto into the GPSO, told a Japanese writer that his shop in San Bruno worked after 1961 to produce counterfeit Vietnamese currency. The assertion seems plausible. Washington reproduced bank notes of both Saigon and Hanoi for use in agent operations and psychological warfare, according to an American military veteran of psychological warfare.[25]

Life in San Bruno for the Japanese technicians appears to have been comfortable. Iwase remembered that the 'government' rented flats for them. They also commuted by military bus from their residences to the nearby naval base, which fronted San Francisco Bay. Discussing the details of their secret work was, of course, forbidden. Otherwise, they faced no restrictions in their daily lives. The Japanese were free to mingle with local Americans, as Iwase recalled doing, and enjoy themselves. Colonel Yamamoto himself indulged his interest in gems by joining a local rockhounding club in Daly City, located between San Bruno and San Francisco.[26]

Yamamoto knew after the closure of the GPSO in Yokosuka that he would have to leave California and find new work at some point. He called on Iwakuro, who had put him in charge of Noborito's Third Section years before. It appears that the two met at least once some time in the mid-1960s. However, Iwakuro had nothing to offer Yamamoto. Iwakuro recalled the meeting in an interview that he gave in 1967 for a history project:

> … [Yamamoto] was taken along to America and apparently has been at it outside of San Francisco. He came back for the first time after a long time, so we threw a party for him. It had been more than ten years since I had last seen him. He said, 'I'll come back after another two or three years, so please take care of me at that time'. I replied, 'Sorry, but I'm done with counterfeit bank notes.'

In 1967, Yamamoto, still living near San Francisco, enrolled in a course of home study at the Gemological Institute of America (GIA). Although the IJA had sent him to Manchuria in 1931 to survey mineral resources along the Soviet border, Yamamoto admitted in a later interview that he had been a 'complete layman' when it came to assessing gems. In 1968, having earned his graduate gemologist (GG) diploma from the GIA, Yamamoto returned to Japan. His departure probably coincided with the end of his contract in California. A chapter in Noborito's postwar intelligence history had come to a close.[27]

CHAPTER 9

REBUILDING JAPAN

> Apart from the crime of war, I have great confidence and pride in having made copious use of the national budget to produce counterfeit bank notes, something absolutely impossible in peacetime, and then, applying and developing techniques acquired at Noborito, to have been able to develop the Meiwa Gravure Company, which at present has more than 4,000 employees in Japan and abroad.
>
> – Oshima Yasuhiro, veteran of Noborito's Third Section and postwar CEO of the Meiwa Gravure Company[1]

Defeated in the Second World War, Imperial Japan lost its military. Gone were the War Ministry, Army General Staff, the AOAH, and its 10 numbered army technical research institutes, including the ninth one: Noborito. The institute's military officers, technicians, and consultants would no longer research and develop spy gear and special weapons for the Kwantung Army, the Kempeitai, the Nakano School or the Yama Agency. The empire no longer existed.

No longer in the Imperial Japanese Army, veterans of Noborito applied their talents to postwar pursuits. They contributed to Japan's recovery from defeat and the nation's emergence by the end of the Cold War as the world's second economic power. Some top officers took senior positions in affiliated corporations that had been working for Noborito on various wartime projects. Many of the technical officers likely would have gone into industry after university had it not been for Imperial Japan's final 15 years of war in Asia and the Pacific, starting in 1931 with the Manchurian Incident and ending in 1945 with the surrender in Tokyo Bay aboard the USS *Missouri*. Noborito's numerous consultants and contract researchers no longer worked for Noborito but retained their academic or corporate positions.

A few joined Japan's postwar defence establishment. Others remained influential while working behind the scenes.

General Officers Take New Positions

Lieutenant General Arisue Seizo kept his hand in intelligence and military affairs after the war. The Imperial Japanese Army's last intelligence chief started his postwar work in developing connections before the surrender with General MacArthur's men at Atsugi and Yokohama. He attached his Arisue Agency to GHQ/SCAP's G-2, brokering intelligence to the Americans. He remained an advisor to the US military in Japan until 1956. He further maintained his American connections over the years, visiting Major General Willoughby in trips to the United States in 1959 and 1966. In 1965, with the founding of the Japan Council of the World Veterans Federation (WVF), Arisue assumed the post of president under the new organisation's chairman, former Finance Minister Kaya Okinori. In 1970, Arisue was elected chairman of the influential Japan Veterans League and represented Japan that year at the 1965 WVF conference in Lausanne, Switzerland, where as part of his speech he read to the assembly a message from Prime Minister Sato Eisaku. In 1974, Arisue oversaw an exchange agreement with the Korea Veterans Association to develop ties to ROK military veterans. He led the following year a delegation of 102 members to South Korea to participate in a ROK military memorial ceremony. Late in life, he penned two memoirs, one of them regarding his actions in Occupied Japan as chief of the Arisue Agency.[2]

Lieutenant General Iwakuro Hideo, active in the establishment of the Nakano School and the Noborito Research Institute, had an extraordinary postwar career. After years of brokering intelligence behind the scenes to the Americans at the head of his own Iwakuro Agency, he became a public intellectual. Seemingly unemployed for 20 years, Iwakuro in 1965 joined the astrophysicist Araki Toshima in founding Kyoto Sangyo University and directing the new university's Institute for World Affairs. He wrote a 1956 account of his role as a regimental commander in the Japanese Army's conquest of British Malaya and Singapore, a general treatise on war in 1967, and a posthumous work in 1971 on his perception of a world evolving from an 'age of science' to an 'age of man'. Iwakuro also engaged Western intellectuals, publishing his 1968 dialogue with the British historian Arnold J. Toynbee on global prospects in the next century and his 1969 talk with the American futurist Herman Kahn on the prospects for science and religion.[3]

Lieutenant General Shinoda Ryo, who had directed the Noborito Research Institute in war, applied his connections and expertise to industry. A man whom a subordinate remembered as 'a top-notch scholar of fibres',[4] Shinoda after the war joined Tomoegawa Paper, one of the companies that had supported Noborito's counterfeiting operations. He rose from laboratory director to company chairman before retiring to a consultant's role. Shinoda also played leading roles in Japanese industry, heading the Society of Fibre Science and Technology, Japan and serving as director of the Japan Technical Association of the Pulp and Paper Industry.[5] He enjoyed haiku poetry and golf in his spare time.[6]

Section, Group Chiefs Find New Roles

The former directors of Noborito's three major sections and some of their subordinate group chiefs found work in their particular fields.

First Section

Leading figures from Noborito's First Section, whose projects included bombing balloons and the death ray, went to work in companies affiliated in the war with Noborito projects. Some resumed their military careers by joining Japan's postwar Defence Agency (DA).

Major General Kusaba Sueki, director of First Section, had overseen Noborito's unsuccessful Ku-go project to develop a death ray. After the war, he became a managing director at Japan High-Frequency, a key company in that project. Dr Sasada Sukesaburo, who had overseen animal experiments in the death ray project as chief of Group 3, also worked for a time at the same company.[7] He later became president of Nihon Dempa Kogyo (NDK), whose products include ultrasonic transducers for medical uses, and served as a director in several other Japanese corporations.[8]

Major Takeda Teruhiko, chief of Group 1, who had worked to develop balloons to bomb the United States, was one of the relatively few Noborito veterans to resume a postwar military career. Takeda became chief of the radar section in Department 4 of the Defence Agency Central Technical Institute and worked there 'for a time' under Lieutenant Colonel Niizuma Seiichi.[9]

Second Section

Veterans of Noborito's Second Section also applied their talents to civilian fields in the postwar era.

Colonel Yamada Sakura, who directed as chief of Second Section the development of tools ranging from poisons to secret writing, became after the war a captain of industry. After heading the laboratories of the Kawaguchi Rubber Works Company, which grew into the international company today known as Lonseal, he rose to the rank of managing director in 1962 before retiring to a consultant's role in 1970. Yamada also served as a manager or consultant in several other companies and industry associations and became a leader in Japan's polyvinyl chloride industry. The recipient in 1942 of the Order of the Rising Sun, 3rd Class, for his wartime achievements in military technology, Yamada received in 1970 a second honour from Emperor Hirohito, the Blue Ribbon Medal, for his contributions to Japanese industry. When not working, he enjoyed gardening and reading.[10]

Group chiefs who worked under Colonel Yamada in Second Section also succeeded in postwar careers.

Major Ban Shigeo, who as chief of Second Section's Group 1 had directed the development of spy gear and materials for the Kempeitai and other IJA units, and had worked to develop poisons, became in Occupied Japan the technical director of the Taimei Chemicals Company in Nagano Prefecture and worked for the United States as the Japanese chief of the GPSO chemical section inside the naval base at Yokosuka. Ban rose at Taimei Chemicals to the rank of vice president in 1974, became a non-executive director in 1980, and retired from the company in 1991. He earned in his postwar career commendations and awards from the Minister of International Trade and Industry (MITI), the Director General of the MITI Agency of Industrial Science and Technology, and elsewhere. A wartime subordinate, Captain Arikawa Shunichi, went from working at Noborito on secret inks to a postwar career at the Toyo Ink Manufacturing Company. There, he headed the company's chemical conversion department and served as an advisor following his retirement.[11]

Major Maruyama Masao, who had developed as head of Second Section's Group 5 technology for microdots, miniature cameras, and other spy gear, worked for two years after the war in the Tokyo District Public Prosecutors Office before resigning and joining Colonel Yamamoto Kenzo at the GPSO in Yokosuka, then in the United States. Following his return to Japan, Maruyama worked for Yamamoto in his former chief's gem authentication venture.[12]

Major Kuba Noboru, who had directed as chief of Second Section's Group 7 the development of cattle plague and other livestock diseases as biological weapons, worked in veterinary science as a researcher affiliated with the Institute for Comprehensive Medical Science at Fujita-Gakuen Health University. In 1990, he expressed his appreciation to wartime colleagues in a 14-page outline of Group 7's efforts to make a weapon of rinderpest, praising those who had been able to 'complete in a short period of time' a way to develop rinderpest as a weapon to 'attack enemy countries and annihilate them.'[13]

Third Section

Veterans of Noborito's counterfeiting and forgery operations, talented technicians, had varied careers. Some worked for the Americans in the GPSO. Others had careers after the war in conventional government and corporate printing outfits. Some joined, or returned to, the government's Printing Bureau. Others applied their skills in leading corporations, including Toppan Printing and Dai Nippon Printing.[14]

Colonel Yamamoto Kenzo, who had directed as chief of Third Section the production of counterfeit bank notes and forged documents during the war in order to undermine China's economy and support agent operations in China, the Soviet Far East, British India, and elsewhere, turned a hobby of his into a new career. Knowing that his contract with the US government was coming to an end in the late 1960s and unable to find a Japanese employer for his special skills, Yamamoto turned to gems. He had conducted mineral surveys as an IJA officer in Manchukuo along the Soviet border in the early 1930s. He made a hobby of gems and minerals as an amateur 'rockhound' in California, joining a local club of mineral enthusiasts near San Bruno. Yamamoto also began learning how to cut gems. In 1967, he enrolled in a course of home study in gemology with the GIA and earned the institute's Graduate Gemologist (GG) degree the following year before returning to Japan. He then worked as a consultant for the Gemological Association of Japan. With the approval of GIA executives, Yamamoto established in 1971 with a Japanese former GIA staffer the Association of Japan Gem Trust (AGT) to oversee GIA courses in Japan and served as board chairman.[15]

After years of counterfeiting in the Second World War and Cold War, Yamamoto turned to authenticating and grading gems. Looking back on his life, he must have felt that it was somehow unreal. Such sentiment is reflected in his starting his memoir with two lines in Japanese translation of a classical Chinese poem by the Tang poet Bai Juyi:

往事渺茫	The past is distant and indistinct
都て夢に似たり	It all seems like a dream[16]

Major Ito Kakutaro, who as director of Third Section's North Group had directed the development of paper for various projects, returned after the war to civilian industry. A graduate of Tohoku Imperial University's Chemistry Department and an employee of the Oji Paper Company before conscription set him on a path to Noborito, Ito later took over his father's paper business, Mantsune, in Aichi Prefecture. A pillar of the community, Ito was a member of the Nagoya Chamber of Commerce and Industry, a member of the city's Rotary Club, and board chairman of the prefecture's paper trade association. He received in 1972 the Blue Ribbon Medal for his contributions to Japan's paper industry.

Oshima Yasuhiro, a technician subordinate to Ito in North Group, also contributed to postwar Japanese industry. Ito, reaching out to his wartime subordinate, invited Oshima to Aichi Prefecture, and arranged for him to join the Meiwa Printing Company. Oshima in time became chairman of the company, now known as Meiwa Gravure. His proudest postwar moment was receiving from Emperor Hirohito in 1975 the Purple Ribbon Medal for technical innovation.[17] Oshima recalled years later that, as a young graduate of the machinery section of his technical high school, he had assembled, improved, and designed at Noborito various machines for printing, paper making, and other processes. He recalled that his experiences and skills had proved 'very useful' when he established his own company.[18]

Major Kawahara Hiroma, who as director of Third Section's South Group oversaw plate making and printing, had been working at the government's Printing Bureau before his transfer to Noborito. After the war, he joined Dainichi Seika Colour & Chemicals Manufacturing Co., Ltd., before leaving to establish and run his own company: Taiyo Ink Manufacturing Co., Ltd.[19]

Contract Consultants

Noborito's wartime contractors, many of whom were leading lights in Japanese academic and government institutions, continued in the postwar years to advance in their careers.

Dr Okabe Kinjiro, awarded in 1944 the Order of Culture for his scientific accomplishments, continued after the war in his career. A professor at Osaka University's Institute of Scientific and Industrial Research from 1941, he directed the institute from 1951 to 1955. He became a 'driving force' there in the development of the microwave oven as a popular consumer good. After mandatory retirement from the institute in 1955, he became the following year a professor at Kinki University, rising to direct in 1966 the Faculty of Engineering at Kinki University Kyushu. He was awarded the Order of the Sacred Treasure, First Class, in 1969 for his contributions to science.[20]

Dr Arakawa Hidetoshi continued in meteorology after the war. In 1964, he became chief of the Fukuoka District Meteorological Observatory. In 1966, he rose to direct the Japan Meteorological Agency's Meteorological Research Institute. Mandatory retirement in 1968, he became a professor at Tokai University, where he worked until 1983. He received in 1972 the Fujiwara Award from the Meteorological Society of Japan for his career contributions to his field. He was awarded the Order of the Sacred Treasure.[21]

From Wartime Biological Weapons to Postwar Japanese and Korean Developments

Two of the OGGC institute's key members, one Japanese, the other Korean, took separate paths after the war. Both succeeded in their respective careers. Before going their separate ways, the two appear to have laid the groundwork for the eradication of cattle plague in Africa.

In June 1946, Dr Nakamura Junji returned with US permission to Japan. He was now free to establish his envisioned Laboratory of Comparative Pathology, What was behind his plan? First, he admitted in his memoir that he had lacked the confidence to try his hand in another field. Second, he had no reason to expect that the MAF Animal Infectious Disease Research Laboratory, the only Japanese facility comparable to the OGGC institute, would provide positions for him or his wartime colleagues. Third, he was hopeful that he could make a go of his new laboratory by producing and selling vaccines. Fourth, he expected that he could count on the help of his OGGC colleagues, who would likely find suitable employment nowhere else.

Dr Nakamura established the LCP that year in 'a corner' of the institute of the Japan Veterinary Medical Association in Tachikawa, at the site of the former Army Veterinary Materials Warehouse. Several wartime colleagues who joined the LCP at its start included Hotta Tokuro and Iwasa Takeshi, both of whom had worked under Nakamura on the Noborito BW project.

Nakamura continued the virus research that he had conducted earlier at the OGGC Livestock Hygiene Research Institute and looked for commercial opportunities. An early success was LCP's producing from the start of 1947 a rabies vaccine for the Japanese government and GHQ/SCAP to use to counter an epidemic of the disease among dogs in postwar Japan. That same year, the LCP and the institute of the Japan Veterinary Medical Association merged to become a MAF non-profit association, the Nippon Institute for Biological Science (NIBS). Nakamura became director of the Research Department and a member of the new organisation's board of directors. The following year, NIBS joined hands with MAF, the University of Tokyo, and the Kitasato Institute to produce a vaccine against Japanese encephalitis in horses. The increasingly successful organisation developed in the course of the Occupation products against such animal scourges as rinderpest, hog cholera, and Newcastle disease.

Nakamura remained active in NIBS under its reorganisation in 1960 as two entities: a research body functioning under the MAF and the Ministry of Education (MOE) and an associated for-profit company, Nisseiken, to sell NIB's livestock vaccines and diagnostic agents. In the new structure, Nakamura acted as standing director for research and as senior researcher. Ochi Yuichi, wartime research director at the OGGC Livestock Hygiene Research Institute, worked as a director and senior researcher with Nakamura.[22]

Over the years, Dr Nakamura became a leader and representative of Japanese veterinary science. In 1961, he became chairman of the Japanese Society for Virology. A frequent Japanese participant in international events, he represented Japan at the US-Japan Cooperative Medical Science Programme Conference, held in October 1965 by the US National Institutes of Health (NIH) in Honolulu, Hawaii, and was a member that year of the conference's viral disease committee. For his efforts, Dr Nakamura received a great deal of recognition, including the Agriculture Minister's Award in 1955 for his virus work, the Purple Ribbon Medal in 1965 for rinderpest vaccine development, and the Science and Technology Agency Award in 1970 for developing purification technology for a Japanese encephalitis vaccine.[23]

In January 1947, two months after crossing the 38th parallel into North Korea and going to Pyongyang, Dr Kim Jong Hui received a summons to the NKPPC. There he met Chairman Kim Il Sung, the North Korean leader. Dr Kim recalled years later that Chairman Kim spoke to him of his ambition to develop the nation's livestock industry, establish a livestock hygiene service and epizootic prevention centres, and produce veterinary and anti-epizootic medicines. A few days later, Dr Kim received appointments as director of the Livestock Hygiene Research Institute and deputy head of the Faculty of Agriculture and Forestry at Kim Il Sung University. He received 'a large sum of money' to develop veterinary science in northern Korea. The institute took shape at the foot of Mount Taesong, on the northeast outskirts of Pyongyang. In May 1948, the construction site received a visit from Kim Il Sung. Under Dr Kim's leadership, the institute produced vaccines for cattle plague and hog cholera. The two epizootic diseases, dangerous to livestock, by design or happenstance had been the two focal points of Noborito's wartime BW efforts at the OGGC Livestock Hygiene Research Institute.[24]

While Dr Nakamura built his postwar career in Japan, his 'favourite apprentice,' Dr Kim Jong Hui, became a pillar of the livestock industry and veterinary science in North Korea. Recognition of his talent from on high protected Dr Kim from persecution. In a speech given at a national gathering of intellectuals in 1999, he stated that Kim Il Sung had overlooked his having 'served in an exploitative society,' a seeming reference to his having worked for the Japanese at the OGGC Livestock Hygiene Research Institute, to make him an instructor at Kim Il Sung University and director of the State Livestock Hygiene Research Institute. Furthermore, Dr Kim stated that Kim Il Sung had 'protected' him and other intellectuals with disadvantageous backgrounds 'to the end' from 'narrow-minded factional elements.'

Dr Kim in the course of his career received the Kim Il Sung Prize and the titles of academician and professor. His children and grandchildren benefited from his success. His son Kim Sun, born in Fusan five months before Korea's liberation in August 1945, graduated in Pyongyang from the prestigious University of Science, conducted livestock research, earned a doctoral degree, and rose to the position of vice president in the Biology Branch of the State Academy of Sciences.[25]

Nakamura and Kim went their separate ways after the Second World War, but the work they did together at the OGGC Livestock Hygiene Research Institute proved of great benefit to Africa. At the 1948 Conference of the United Nations Food and Agriculture Organisation (FAO) in Nairobi, Kenya, Republic of China (ROC) representative Cheng Shao-chiung reported that he had obtained highly effective results in protecting cattle in China against rinderpest by inoculating them with stocks of the rinderpest virus that Dr Nakamura had left behind in the course of the wartime work that he and his assistant Isogai Seigo had conducted. The FAO promoted in the 1950s the use of Dr Nakamura's rinderpest vaccines to protect cattle herds in Africa. Dr Nakamura himself spent four months touring Africa in 1957 and another three months in Egypt in 1961 to promote the inoculation of cattle against rinderpest. His efforts played a large role in the eventual eradication of rinderpest from the world, which the FAO declared in 2011.[26]

As for Dr Kim, an article in a Pyongyang monthly magazine reported his having received one summer day 'in the 1960s' a letter from 'a Japanese academic' informing him that previous work of his had protected cattle in Africa against rinderpest.[27] The article omitted for reasons of propaganda the key detail that Dr Kim had worked on the rinderpest vaccine at the OGGC Livestock Hygiene Research Institute as Dr Nakamura's 'favourite apprentice.' The letter, perhaps written by Dr Nakamura himself, suggests that Dr Kim and, by extension, other Korean staff, contributed to the development of an effective rinderpest vaccine.

Common and Uncommon Lives

Many senior Japanese military officers and civilian technicians had great success in their postwar careers, but most of the men and women who worked at Noborito or affiliated organisations lived modest lives. Many veterans remained near where they had worked in the war. The men in many cases found work at some local organisation, married, enjoyed some success in their careers, and lived ordinary lives as family men. One example was that of Miyamoto Yoshiro. A former employee at Noborito's Fourth Section, he found work after the war at the municipal waterworks bureau in nearby Kawasaki.[28]

In the society of that time, most of the young women who had some connection to Noborito in making the bombing balloons or labouring on some other project, followed the path taken by most women of their generation. They married, became mothers, managed their households, and found outside work. Seki Koto, the typist from Second Section, married and, as Mrs Kobayashi, became a mother and grandmother.[29]

Some men who had worked on sensitive wartime projects lived in fear of exposure or prosecution for their work at Noborito. In the Occupation period, American intelligence officers and Japanese police interrogated many of the veterans. The Americans, seeking new military technology, engaged in what Colonel Yamamoto had described as 'give and take,' in which the Americans kept Noborito's men in Occupied Japan from going on trial as war criminals in exchange for their cooperation. It was this American desire for Japanese technical intelligence, according to Major Ban, that resulted in nobody from Noborito appearing before the Tokyo War Crimes Tribunal.[30]

Fear of prosecution led one man to abandon Japan and live his postwar life in disguise overseas. A researcher at Noborito's Second Section working on the destruction of plants, he was on assignment to the Kwantung Army in Hsinking when the Soviet Red Army invaded Manchukuo near the end of the war. He managed to make his way by train south to Shanghai. Fearing the interrogation that awaited him if he surrendered to the Chinese authorities and entered a camp for repatriation, he abandoned the idea of returning to Japan, made his way to Hong Kong, and lived there under a Chinese name.[31]

The Japanese police grilled many men from Noborito's Second Section in connection to a shocking crime that took place one day in January 1948. A man posing as a municipal quarantine officer entered near closing time a branch of the Teikoku Bank (also known in abbreviated Japanese as Teigin) in Tokyo's Shiinamachi area. Wearing an official armband and presenting a business card that identified him as an employee of the Ministry of Health and Welfare, he announced to the employees that there had been a nearby outbreak of dysentery and that he required that everyone present drink some medicine before Occupation authorities arrived to disinfect the area. The 16 persons present drank the liquid, a poison that acted with a delay. The criminal escaped with some cash following his deception, which killed 12 of the 16 victims. When police discovered that the murderer had administered a hydrocyanic acid compound with a delayed effect, an uncommon poison, they began interrogating former members of Noborito as well as other military units.

Former members of Second Section, which had developed acetone cyanohydrin, whose delayed action made it a tool for assassination, came under intense scrutiny. Captain Takiwaki Shigenobu, who had worked under Major Hijikata Hiroshi in Second Section's Group 3 to develop acetone cyanohydrin and other poisons[32], told a Japanese writer years later that the Japanese authorities had subjected him to 'severe interrogation,' as they had many of his colleagues, in the course of their investigation. Reflecting years later on his experience, Takiwaki said that he had no 'personal feelings' regarding the poisons that he had developed in the war on the orders of his military superiors but described his life after the war as a 'thorny path.'[33]

In the early 1960s, police investigated a series of counterfeiting cases uncovered throughout Japan. Known as the Chi-37 Incident, it was the nation's most serious counterfeiting case. The police, aware of by that time of Noborito's wartime counterfeiting activities, interrogated veterans of Third Section but found none of them to be guilty. The case was closed, unsolved, in 1973.

Some Noborito veterans felt bitter about their past. Sergeant Major Sugita Masami, who had worked in Noborito's Fourth Section, responsible for testing items developed at Noborito, told a reporter that some Second Section personnel involved in development of poisons and other biological weapons stayed away from postwar reunions. Sugita himself, who described to a reporter his section's work on explosives, grew evasive and stopped talking when questioned on projects involving poison gas and biological warfare. He assessed his own experience in bitter terms: 'For me, my time at the Noborito Research Institute was a complete waste.'[34]

Passing into History

Time flows without cease, sweeping all those in its stream from this transient world into eternity.

As the postwar years passed, the ranks of those associated with Noborito grew ever thinner. Lieutenant General Iwakuro, the founding father of the Army's institute for spy gear and special weapons, passed away in 1970. Lieutenant General Shinoda, the institute's director, died in 1979. Illness took chief counterfeiter Colonel Yamamoto in 1988. Lieutenant General Arisue, intelligence broker, passed away in 1992.

Major Ban, Shinoda's closest assistant, passed away in 1993. Dr Kim, the scientist who had worked both for the Japanese Empire and the DPRK, died in 1996.[35]

Time also brought changes to the former 9th Army Technical Research Institute. The US Army took over the Noborito site in October 1945. Occupation authorities allowed Keio University, the Kitasato Institute, and Tomoegawa Paper to use facilities on the site until 1950, when Meiji University acquired most of the land to relocate its Faculty of Agriculture to Noborito on what became the university's Ikuta Campus.[36] Noborito's main administrative building, earlier the main building of the Japan Higher Colonisation School, survived into the 1960s as a library. Over time, administrators tore down the old, decaying structures. Building 36, the last building, exists today as a museum dedicated to the history of the Noborito Research Institute.

CHAPTER 10

REMEMBERING NOBORITO

> The Noborito Research Institute's research and the weapons and materials developed there include items that in some cases pose major problems in terms of humanity or international law. However, I believe that we need to face this dark side of the war and pass on calmly and dispassionately to future generations the true nature of war and some of the actions that the Japanese military carried out.
>
> – Yamada Akira, curator, Defunct Imperial Japanese Army Noborito Laboratory Museum for Education in Peace[1]

After the Second World War, former members of Noborito gathered together to see old friends and recall old times. As time passed, some veterans organised to erect a monument to the past. Veterans and local activists also called for the preservation of Noborito's remaining structures and the presentation of its history. Their efforts resulted in the creation of Japan's only museum dedicated to clandestine operations of Japan's military.

Reunions and Remembrances

Men and women at Noborito after the war wished to see their old colleagues and friends from time to time. As soldiers, sailors, and civilians the world over have joined veteran associations and shared memories with comrades in arms, so did Japanese veterans. Many IJA officers joined the Kaikosha. Former IJN officers became members of the Suikosha. Instructors and graduates of the Nakano School formed their own association.[2]

Noborito veterans established their own groups. After the war, some met and put together lists of names and contact information. In 1946, some of the younger employees living near Noborito gathered to ring in the new year. In 1950, former members of Third Section met and put together a contact list. In 1961, Noborito veterans involved with the GPSO circulated a name list among themselves. The following year, members of Fourth Section had their first meeting. In 1963, some veterans of Second and Fourth Sections compiled their own name list. Third Section's Oshima Yasuhiro recalled that some 20 to 30 members of his section would gather together each year in August, the month that the fighting had ended.[3]

In 1982, veterans formed an association for the entire institute, the Tokenkai, a Japanese abbreviation for the Noborito Research Institute Association. The first reunion, organised by Captain Nakamoto Toshiichiro, Major Takano Yasuaki, and several other veterans, was held that year at the Hotel Nanpuso, a popular inn of the Hakone Yumoto hot springs district.[4] Members of the Tokenkai decided to erect a commemorative stone marker in a corner of what had become Meiji University's campus in Ikuta, Kanagawa Prefecture. In April 1989, the Tokenkai placed the marker near Noborito's former Yagokoro Shrine. Members gathered in November that year for the formal ceremony, with Major Ban Shigeo unveiling the monument.

Carved into the monument's front were the characters marking the site as that of the former Noborito Research Institute. On the back was carved in Japanese a haiku poem:

すぎし日は　Bygone days
この丘に立ち　Standing on this hill
めぐり逢う　We meet again

For Noborito's veterans, the words expressed a sense of liberation from the secrets that had bound them for decades. The third line's reference to meeting again implied that the time had come when it was possible to gather and speak freely with those from other sections as well as colleagues from their own about what they had done in the war at Noborito.[5]

Also carved into the back of the stone was the Tokenkai's intended date for erecting the marker: October 1988. The delay in the monument's actual placement was due to Emperor Hirohito's failing health. The previous month, the government had announced that he was in critical condition. The announcement had stirred up students at Meiji University. The campus remained unsettled even after his passing in January 1989. Noborito veterans waited until the students went on spring break in April to place their marker.[6]

As time passed, Noborito's surviving veterans felt less constrained and better able to share their secrets with the public. The passing of Japan's wartime emperor and military leaders made Noborito's surviving veterans feel less constrained and better able to share their secrets with the public. Or at least some secrets. Ban revealed many details in a 1980 article on Noborito but wrote only that the counterfeiting Third Section had engaged in research 'related to printing.' Four years later, Colonel Yamamoto published a detailed account of Third Section's counterfeiting operations but ended the story abruptly in 1945. His book contained not a word about his postwar cooperation with the Americans. Even Ban's posthumous 2001 history, the most complete firsthand account of Noborito's secret history, including what he knew and the recollections of key colleagues, left much in the dark. He had cut from an early draft details of a key encounter in 1948 with US intelligence officers that led to his joining the GPSO, unable to the end of his life to publish his secrets of the postwar years. Noborito veterans who worked with Washington, sworn to secrecy, took nearly all their secrets to the grave.[7]

Preserving the Past, Liberating Veterans

As the years passed, members of the Tokenkai and local citizens lobbied Meiji University to preserve what remained of the Noborito Research Institute and to establish a museum to tell its story. In a letter dated 30 October 2005, Yamada Genzo, a technician from Noborito's First Section, submitted as representative of the Tokenkai to Meiji University President Naya Hiromi a letter calling on the university to preserve Noborito's remaining structures and establish a museum. He laid out the purpose: 'We worked at that Army Noborito Research Institute. There, directly under the Army General Staff, there took place research and manufacturing of weapons for clandestine warfare.' Yamada explained that First Section was for physical weapons, Second Section for biological and chemical ones, Third Section for economic covert operations, and Fourth Section for the development and manufacturing of those weapons that could be mass produced. Yamada's letter continued: 'We consider that, even though it was at that time an institute for clandestine warfare, the facts should be preserved as they are and stand the trial of history.' Yamada closed the letter by writing that Tokenkai members were willing to provide materials from that time and to cooperate in establishing the proposed museum.[8]

Few physical traces of the Noborito Research Institute remained by the time President Naya received the Tokenkai's letter. Meiji University administrators over the years had demolished the aging wartime structures on its Ikuta Campus and, in some cases, replaced them with modern buildings. Several of the old buildings were still standing when the Tokenkai had placed in 1989 its memorial near the shrine. In 1999, Meiji University took down the structure known as Building 44, which had served Second Section in developing weapons for biological and chemical operations. In 2009 came down Building 26, which Third Section had used to store its counterfeit bank notes. Building 5, Third Section's printing plant, was razed in 2011. The space once occupied by Buildings 5 and 26 became a car park.[9]

Meiji University preserved a single structure: Building 36. Noborito's Second Section had used the modern building – built with reinforced concrete, in contrast to most of the other buildings, wooden structures topped with tile roofs – as a space to experiment and develop biological weapons for attacking rice, wheat, and other crops. After the war, Meiji University's Faculty of Agriculture had used it as laboratory space until 2009.[10]

It was this building that became the museum to preserve the history of the Noborito Research Institute. On 29 March 2010, Meiji University hosted a ceremony to inaugurate the museum. Noborito veterans, members of the Kawasaki government, an area citizen's group that had lobbied for preserving Noborito's history, President Naya, and other members of Meiji University were among the event's 120 participants. Representatives standing on a red carpet placed outside the building cut the ceremonial ribbon for Meiji University's new facility. President Naya in his speech expressed hope that the museum would serve as a site for education in history, peace, and science. On 7 April that year, the building opened to the general public. Various Japanese media and a South Korean television news team were on hand to report the opening. The museum has received more than 90,000 visits since then.[11]

The museum's five exhibition rooms showcase various aspects of the Japanese Army's Noborito Research Institute. The first room contains an overview of Noborito's establishment and expansion. The second room covers the activities of First Section, including exhibits on the institute's development of bombing balloons and its work on a death ray. The third room tells the story of Second Section, with its biological weapons, poisons, and spy gear. The fourth room holds information on Third Section's counterfeiting of Chinese bank notes and fabrication of foreign documents. The fifth room traces the withdrawal of Noborito units to more secure locations as part of the military's plans for its decisive military campaign on Japanese soil and relates what happened after the war ended.

In addition to its exhibition rooms, the museum serves as a centre of exhibitions, conferences, and events that teach the history of the Noborito Research Institute, the Imperial Japanese Army, and Japan's 15-Year War (1931-1945), waged across much of Continental Asia and the Pacific Ocean. Past exhibitions have presented such topics as 'The Decisive Battle on Japanese Soil and Secret Warfare' (2013–2014), 'Science, Technology, and Civilian Mobilisation in War' (2017–2018), 'The Teikoku Bank Incident and the Noborito Research Institute' (2018–2019), and the 'Top Secret "Yama Agency" and the Noborito Research Institute' (2022–2023). Professional staff produce the annual 'Museum Report' and the more frequent 'Museum News' series. In addition, the museum communicates through its website and the online platforms Twitter, Facebook, Instagram, and YouTube.[12]

The Museum's tone is purposeful. Visitors learn details of Noborito's projects, technical aspects of the special weapons and spy gear, particulars of the institute's organisation, and information on such difficult wartime history as the people killed in experiments to develop Second Section's poisons. Such an approach reflects the thinking behind the museum's establishment, that of 'facing what could be called the "dark parts" of war' in showing the 'true nature of war and some of the actions' of the Japanese Army.[13]

Oshima Yasuhiro of Third Section expressed what has become the museum's purpose in a note of congratulations, dated 20 March 2010, following the inauguration of the museum. At 88 years of age, he reflected in his letter: 'What is clear to me now is that what we were testing and producing at Noborito were the kinds of weapons that appeared in the science fiction published from the time when we were children. The bombing balloons were done at First Section, the poison gas, bacteria, and such at Second Section. War is awful! We no longer need such weapons. Let us create a cheerful society.'[14] The museum's intention, by exhibiting Noborito's secret wartime history, is to give visitors a greater appreciation for peace.

The museum conferences and exhibitions are also meant to impress on Japanese visitors that their country was an aggressor as well as a victim in the empire's final 15 years of war, an antidote to Japanese education and media accounts that stress Japanese suffering. Yamada Akira, the museum's curator, sees the museum helping to 'bridge the gap' in Japanese memories of the war and those in 'neighbouring countries,' serving as a foundation on which to build 'new relations' between Japan and Asian nations. According to Yamada, another aim of the museum is to encourage reflection on how to use technology for peaceful ends and to consider in the light of Japanese history Tokyo's loosened restraints in recent years on military action.[15]

Apart from such aims, the museum has also brought a sense of relief to Noborito veterans long burdened under the weight of their secrets. Noborito historian Watanabe Kenji recalled the case of one particular visitor to the museum. He had worked in the war for Noborito's Third Section. He continued his secret career after the war in working for years with the Americans at the GPSO in Yokosuka, then California. Watanabe described the veteran as someone living his life unable to speak of his past. Visiting the museum, the retired counterfeiter saw a detailed exhibition on Noborito's wartime counterfeiting and postwar work for the Americans. His eyes brimming with tears, the veteran told Watanabe that he felt a great sense of relief that he no longer had to keep silent. As veterans had told Watanabe at a 1998 Tokenkai reunion, their pay and working conditions had been good, but they had done some bad things. They had left Noborito in 1945, promising their superiors that they would take their secrets to the grave. Some had done the same in parting company with the Americans.[16]

What Remains at Noborito

Outside the museum, the Ikuta Campus of Meiji University still holds a few reminders of the site's secret past. Visitors to the campus will find inside the main entrance the imposing stone monument, nearly three metres high and a metre wide, that Lieutenant General Shinoda erected in March 1943 to appease the souls of the animals who died at Noborito in laboratory experiments, and the people in China who died in his institute's poison experiments.[17]

In the northwest part of the campus, nestled under some trees on a small rise, stands what was once the Yagokoro Shrine. Shinoda built Noborito's Shinto shrine – with traditional *torii* gate, purification fountain, and main sanctuary –as a branch of the ASRI's shrine in Tokyo. Technician Kitazawa Ryuji recalled that the main shrine at the ASRI was dedicated to Yagokoro Omoikane no Kami, the Japanese god of invention. The god came to dwell at Noborito's branch shrine as well. It is said that personnel who died at Noborito were enshrined there.[18] In dying at Noborito, they became gods of research, as Japanese soldiers who died in battle in Imperial Japan's wars are enshrined as gods of war in Tokyo at Yasukuni Shrine.

Traces of Noborito in the United States

A few traces attesting to the history of the Noborito Research Institute exist in the United States. Northeast of Bly, Oregon, stands the Mitchell Monument in memory of the only Americans to die in the continental United States as a result of enemy action in the Second World War. A bronze plaque attached to the stone monument lists the name of Mrs Mitchell and the five children who perished on a picnic. In Dayton, Ohio, the National Museum of the United States Air Force includes photographs and other materials related to the bombing balloons. The archives of the Smithsonian Institution's National Air and Space Museum in Washington, DC, hold three boxes of documents and photographs that curator Robert C. Mikesh gathered to write his pioneering book on the bombing balloons. The National Archives at College Park, Maryland, and other archives around the United States hold various documents pertaining to the bombing balloons. In unpopulated desert and lonely woods, debris from bombing balloons likely remain undiscovered to this day.

Traces of Noborito on the Korean Peninsula

The Korean port city known in the Japanese imperial era as Fusan, after in the Cold War as Pusan, and today as Busan, retains something of a trace of the Noborito legacy. In the war, the Noborito Research Institute had rinderpest developed there as a weapon to destroy enemy cattle herds. After the war ended in 1945 and authorities in Seoul established in 1948 the Republic of Korea as the sole government for the divided peninsula, the ROK government maintained the Pusan facility. Added to the Rural Development Administration in 1962, the organisation was renamed the National Veterinary Research Institute in 1994 and became part of the Ministry of Agriculture and Forestry in 1998.[19]

What of Dr Kim Jong Hui's State Livestock Hygiene Research Institute, built before the Korean War at the foot of Mt. Taesong in the rival Democratic People's Republic of Korea? For that matter, what knowledge of rinderpest as a weapon remained in Korean hands after the Japanese researchers departed the peninsula? Are there today in either or both of the two Koreas biological weapons or know-how from the days of the Noborito Research Institute?

Considering Noborito's Place in History

With the passage of time, some of the history of the 9th Army Technical Research Institute – Noborito – has emerged in part from the shadows. Most of the details remain in the dark.

Journalists published soon after the Japanese surrender sensational accounts of bombing balloons and a death ray. Much of the initial reporting was mocking in tone, such as a United Press report of 15 August 1945 that derided the Fu-go project as 'a bizarre attempt to gain aerial equality with the United States.' Major General Kusaba complained that US writers took pains to portray it as worthless. Much of the media reporting was sensational, such as the UP story of 7 October 1945 that Japan had 'perfected' a 'death beam' whose research had begun 'in Noborito in the ninth military technical laboratories' but that American investigators were 'extremely sceptical' of its 'value as a military weapon.'

Japanese reporting proved little better. Lieutenant General Shinoda and Major Ban complained in their overview of Noborito, published in 1977 as a chapter in a book on IJA weapons, that coverage of the institute in 'some' Japanese books and magazines had been nothing more than 'exposés' and that there had been no serious reporting on Noborito's 'technical aspects.'[20]

Missing from such media accounts over the years was a larger perspective. The War Ministry established the ASRI to close Japan's gap in military technology with the Western powers after the First World War. The need to develop spy gear and special weapons to meet military requirements resulted in the 9th Army Technical Research Institute. Noborito was a logical development for Imperial Japan, as was the R&D Branch of the American OSS, Station 12 of the British SOE, and corresponding organisations elsewhere.[21]

The unconventional character of some Noborito weapons, far from bizarre, was the result of Japan's increasingly desperate circumstances in the course of the Second World War. As two writers plainly put it, the Japanese Army worked to develop such innovative, unconventional weapons as the death ray because there was no way that Japan could wage war against the far stronger Allies by conventional means.[22] Inferior in conventional weapons and materials, unable to send aircraft on bombing missions over the United States, the Japanese Army via Noborito was able to turn paper and paste into balloons that bombed North America. Had the Japanese Army kept to the original plan of loading the balloons with biological weapons, the world's first intercontinental weapon could have caused great damage.

Assessing Noborito's Projects

Noborito veterans viewed their work in various ways: as a waste of time, a qualified success, a failure that led to success elsewhere, or simply impressive in terms of technology.

Major General Kusaba, director of the Fu-go project, viewed his bombing balloons as a qualified success. He wrote in a 1961 magazine article that the American technical investigators who reached Japan soon after Japan's surrender made the bombing balloons the focus of their detailed investigations on Noborito. Kusaba claimed that the balloon bombings had succeeded 'to some extent' by causing the enemy much concern. In colourful language, he wrote that the balloons had 'made leaders of the US military tremble!' Kusaba exclaimed that the weapon was 'the Pacific War's version of an intercontinental ballistic missile (ICBM), made in Japan of traditional Japanese paper and konjak!' Lieutenant General Sato Kenryo, director of the IJA Military Affairs Bureau, used the same language in the latter half of the 1960s to describe the bombing balloons: 'They were a primitive version of the ICBMs in which the United States and the Soviet Union are now competing.'[23]

Beyond intercontinental bombing balloons, the Noborito Research Institute succeeded in producing radio direction-finding equipment to break a Soviet spy ring; devices to record in secret the remarks of political dissidents for monitoring and arrest, including a postwar Japanese prime minister; pens that released bacteria to contaminate wells; poisons for assassinating enemies; counterfeit bank notes for economic warfare; and forged passports for operations in enemy territory. In other cases, such as the Ku-go project to develop a death ray, the institute failed to field a weapon or equipment under development.

Like 'Q' in the James Bond movie series, the scientists and technicians of Noborito worked on a wide range of special weapons and spy gear. Unlike the long-suffering 'Q,' whose moments of exasperation at 007's inattention have been comic moments in the Bond movies, Lieutenant General Shinoda directed a real institute tasked with equipping men of the Kwantung Army, Kempeitai, Nakano School, and Yama Agency with the tools of their operations against foreign and domestic enemies of the Japanese Empire. In reality, developers of spy gear and special weapons have harnessed science and technology for projects that, as the museum curator Yamada Akira put it, 'pose major problems in terms of humanity or international law.' Such is the outcome when nations struggle for power in a world without effective international rules or the collective means to enforce them.

Questions

The Japanese Army destroyed incriminating evidence of its special research, with the Fu-go project's bombing balloons and Noborito at the top of Lieutenant Colonel Niizuma's list for destruction, so we will never know many of Noborito's details. Questions abound.

Still missing today from the public record are many details on the biological weapons developed for the bombing balloons. We have some knowledge of the rinderpest virus prepared to attack American cattle, but only the slightest of references to work on hog cholera at Fusan and plant diseases on the island of Tanegashima. What else was under development? Some prisoners in China and Manchukuo died in the testing of Noborito's poisons. Did others die in other projects, such as the development of BW payloads for the bombing balloons? What projects placed Noborito above Unit 731 in the document of 15 August

1945 ordering the destruction of incriminating evidence of 'special research'?

In North Korea, Dr Kim Jong Hui's accomplishments remain unclear. He was the father of livestock science in Pyongyang. Before then he was the 'favourite apprentice' of Dr Nakamura at the OGGC Livestock Hygiene Research Institute in Fusan. Did he take with him to Pyongyang the secret of turning the rinderpest virus into a weapon? With such knowledge, Dr Kim could have pioneered the development of biological warfare in North Korea. Is that why the DPRK leader Kim Il Sung protected him from hostile 'factional elements?'

How many of Noborito's veterans worked for the United States in the Cold War? A little information has emerged on Uncle Sam's hiring some of them to work in Yokosuka and California in support of clandestine operations, but what of other veterans? Washington had a great appetite for technical intelligence.

Wada Kazuo, one of First Section's younger employees, told a Japanese reporter in 1995 that the US Army flew some veterans of the Fu-go project to the United States. If true, who were they? How long did they stay? What did they do? What of Noborito's Second Section? Dr Kim wrote of the US Army plotting to take him and others to the United States. Did the US Army actually send Korean members of the Fusan institute to the United States? If so, what did they do there and where did they do it?[24]

Future Developments

Kusaba wrote in 1961, a time when ICBMs capable of reaching targets across any distance made the bombing balloons of 1945 seem like a tale from a distant past, 'The rapid pace of progress in science is simply astonishing.' Today, the frontier of military science has advanced far beyond where it was when Kusaba made his observation. Scientists and engineers today must be developing spy gear and special weapons far beyond what was possible in the years of the Second World War and the Cold War.

Tokyo, under Washington's wing in the Cold War, favoured industrial expansion over military power in that era. IJA Lieutenant Colonel Fujiwara Iwaichi, who retired from Japan's postwar Ground Self-Defence Force (GSDF) as a lieutenant general after a career that included time as director of the GSDF Intelligence School, wrote in the 1980s that

Japan's postwar military lacked the resources available to the Imperial Japanese Army, including the Nakano School for training 'the necessary personnel for intelligence and clandestine operations' and the Noborito Research Institute 'for the research and development of the required special materials' for such activities.[25]

Japan has been moving to upgrade its military and harness scientific advances as China grows stronger, the Korean Peninsula remains unstable, and America's security guarantee appears less credible than in the decades of the Cold War. Tokyo's upgrading in 2007 of the Defence Agency to the Ministry of Defence (MOD) was one indication of changes in Japan. Innovating to counter China's conventional superiority, the MOD's Acquisition, Technology & Logistics Agency (ATLA) in 2022 allocated funding for development of high-powered microwave (HPM) weapons and set fiscal year 2026 as the goal for a prototype HPM weapon.[26]

The MOD's ATLA appears to be developing a special weapon somewhat resembling Noborito's death ray. What about spy gear and other equipment for clandestine operations? Has Tokyo already established somewhere one or more organisations to supply tools and materials for Japan's developing intelligence community? If not already, is that day not far in the future? It would be no surprise to learn one day of Japanese researchers with top-secret clearances crediting Noborito as the forerunner of their organisation.

NOTES

Preface

1. Ban and Shinoda, '*Noborito*,' 682.
2. Histories of these organisations include John Lisle's *Dirty Tricks Department* (2023) for the OSS Research and Development Branch and Des Turner's *Aston House* (2006) on SOE's Station 12.

Acknowledgements

1. *See* my review of *Rikugun Noborito Kenkyujo no shinjitsu* [The Truth About the Army Noborito Research Institute], by Ban Shigeo, *Studies in Intelligence* (2002), and 'The Japanese Army's Noborito Research Institute,' *International Journal of Intelligence and CounterIntelligence* (2004).

Chapter 1: Building Noborito

1. Iwakuro, *Iwakuro*, 133.
2. Gerald J. Fitzgerald, 'Chemical Warfare and Medical Response During World War I,' *American Journal of Public Health*, April 2008, 612.
3. Theodor Rosebury, 'Some Historical Considerations,' *Bulletin of the Atomic Scientists*, June 1960, 228.
4. Yardley, *Secret*, 31-32, 43-44.
5. Hunt, *American*, 18; Hall, *You're*, 79.
6. Ariga, *Nihon*, 123-126. The term 'special means' could refer to communications intelligence, known in the Japanese Army as 'special intelligence.'

7. AFFE, *Japanese Intelligence Planning Against the USSR*, Washington, DC, Office of the Chief of Military History, Department of the Army, 1955, 8, 33.
8. On the SSAs, IJA field units engaged in intelligence collection and clandestine operations in the Soviet Far East, China, Manchuria, Southeast Asia, and Europe, *see* Hata, *Nihon*, 374-382. On the Nakano School, *see* Mercado, *Shadow*. On the Yama Agency, *see* Ariga, *Nihon*, 80-81, and Iwakuro, 'Junbi,' 17.
9. Ban, *Rikugun*, 165-168, 170, 174.
10. Ban, *Rikugun*, 17; Satake, 'Denpa,' 571.
11. Ban and Shinoda, 'Noborito,' 683; Ban, *Rikugun*, 17-19; Ban, 'Himitsu,' 97.
12. Ban and Shinoda, *Noborito*, 682; Meiji Daigaku Heiwa Kyoiku Noborito Kenkyujo Shiryokan. *Meiji Daigaku Heiwa Kyoiku Noborito Kenkyujo Shiryokan gaidobukku* (hereafter Meiji, *Gaidobukku*), 9; Iwakuro, *Iwakuro*, 32-33, 139.
13. Kusuyama, '*Rikugun*,' 151-52.
14. Nagano/Akaho, *Kokosei*, 99; Shiina, 'Tenji,' 37; Kimura Manabu, 'Zassho tsuzuri: Jittai chokushi shi, kokai ketsui,' *Yomiuri Shimbun*, 25 August 1995.
15. Ban, '*Rikugun*', 20; Kusuyama, '*Rikugun*', 90; Saito, *Boryakusen*, 32.
16. Shiina, "Noborito,' 39. Ban admitted to US Army interrogators in 1946 his 'keen interest' in Yardley's book and also named as references the Lucas book and a German work on chemical analysis, Richard Berg's *Die analysis verwendung von o-oxychinolin ('oxin') und seiner derivate* (1938). *See* CCD, 'Interrogation of the Two Foremost Japanese Secret Writing Chemists,' 10 July 1946, 7, RG 331, box 8554, National Archives at College Park, MD (hereafter NACP). Nagano/Akaho, *Kokosei*, 63; Ban and Shinoda, *Noborito*, 686-87. Director Kusuyama Tadayuki discovered in preparing his 2013 documentary on Noborito that the shelves of the late Major Ban's home were lined with 'spy books.' Ban Kazuko said that her husband had donated many others to the postwar GSDF Intelligence School. She suggested his having instructed students at the Nakano School had motivated him. *See* Kusuyama, '*Rikugun*', 38.
17. Ban and Shinoda, *Noborito*, 683, 687.
18. Ban, *Rikugun*, 26; AOAH. *Rikugun gijutsu kenkyujo no bu/fuhyo: Dai Kyu Rikugun Gijutsu Kenkyujo jininhyo*, 31 August 1945, Japan Centre for Historical Records, National Archives of Japan (hereafter JACAR), Ref. C15010410200.

19. Ban, 'Himitsu,' 101-103; Kimura Manabu, 'Saikin to doku gasu: Yami no bunya, jujika omoku,' *Yomiuri Shimbun*, 30 August 1995; Ban and Shinoda, 'Noborito,' 689; Inaba, *Boryaku*, 30.
20. Ban, *Rikugun*, 57-59.
21. Nakano Koyukai, *Rikugun*, 129; Hatakeyama, *Hiroku*, 108; Saito, *Showa*, 23.
22. Ban and Shinoda, *Noborito*, 685.
23. Saito, *Showa*, 142-48; Saito, 'Zoruge,' 343-344. Kotani Ken sees the origin of Sorge's downfall in the detection of Clausen's transmitter. *See* Kotani, *Nihongun*, 76.
24. Ban, 'Himitsu,' 97; *Koho Kinmu Yoin Yoseijo otsu shu choki dai ichi ki gakusei kyoiku shuryo no ken*, 1939, JACAR, Ref. C01004653900; Saito, *Supai*, 158; Nakano Koyukai, *Rikugun*, 58, 132-33.
25. Kuwahara, 'Nakano,' 249.
26. Ban, *Rikugun*, 34. Ban did not date the organisation chart. Yamada Sakura became director of Second Section from August 1943, which suggests the information is from around that time. *See* Shiina, 'Tenji,' 58.
27. Basic information on the military careers of commissioned military and naval officers comes from Toyama and Joho, *Riku-Kaigun.* Matsuno, 'Nihon,' 23; Hata, *Nihon*, 79; Tsukamoto, 'Rikugun,' 23.
28. GHQ Technical Intelligence Detachment. *Japanese Research on High-Frequency Electric Wave Weapons* (hereafter GHQ/TID, *Japanese*), 24 June 1949, 7, RG 319, box 2119, NACP.
29. Shiina, 'Tenji,' 59; Yamada, *Kagaku heiki*, 1.
30. Yamamoto, *Rikugun*, 245.
31. Shiina, 'Noborito,' 51.
32. Nagano/Akaho, *Kokosei*, 80-81.
33. *Rikugun Kagaku Kenkyujo Shutchojo no meisho oyobi ichi ni kansuru ken*, 1939, JACAR, Ref. C01004565200.

Chapter 2: Developing a Death Ray

1. Ban, *Rikugun* 112.
2. Satake, 'Denpa,' 594.
3. Wells, *War*, 36; Seydewitz and Doberer. *Todesstrahlen*. The next year a French translation was published: *Les Rayons de la mort et autres nouveaux engins de guerre*.

4. '"Death Ray" Is Carried by Shafts of Light,' *Popular Mechanics*, August 1924, 189-190.
5. 'Invisible Death,' *Time*, 21 April 1924, 19.
6. INVENT DEATH RAY THAT 'COULD WIPE OUT ARMY,' Universal Newsreel Volume 2, Release 10, NACP, https://catalog.archives.gov/id/234271475; DEATH RAY SLAUGHTERS BUGS: SAN FRANCISCO, CA, Universal Newsreel Volume 8, Release 461, NACP, https://catalog.archives.gov/id/234272133; DISPLAYS DEATH RAY FOR FEE: SAN DIEGO, CA, Universal Newsreel Volume 8, Release 476, NACP, https://catalog.archives.gov/id/234272148.
7. 'Inventor Reveals Super Death Ray,' UPI, New York, 12 July 1934. A Universal Service report of 10 July 1934, 'Greatest of Inventors Tells Plan to Force Peace,' dateline New York, reported Tesla proposing a '"wall of force" – an invisible screen through which nothing could pass. Tesla reportedly envisioned the wall as 'a beam of concentrated energy' that would 'annihilate airplanes, armies, navies or anything that comes into its path.'
8. Nagase-Reimer and Kawamura, 'Nihon,' 4, 13-14; Eduard Rhein, *Gibt es wirklich todesstrahlen*? The company included it the following year in a collection of German works in Japanese translation, then republished the book in 1942. *See* Matsudaira, *Kindai*. The author referenced Tesla's claim on page 117.
9. 'History of Ultrahigh-Frequency Research,' 3, in Part I, 'High-Frequency Electric Wave Weapons,' in GHQ/TID, *Japanese*, 24 June 1949, RG 319, box 2119, NACP. Satake recalled as influential a press item on electric waves stopping an automobile's engine in Germany. *See* Satake, 'Denpa,' 595. On Yagi's remarks, *see* Nagase-Reimer and Kawamura, 'Nihon,' 3.
10. GHQ/TID, *Japanese*, 3. Noborito technician Yamada Genzo credited Sasada as an expert on the physiological effects of short waves. *See* Ban, *Rikugun*, 114.
11. Ban, *Rikugun*, 111; Satake, 'Denpa', 580.
12. Records of the US Strategic Bombing Survey (Pacific) (hereafter USSBS), Interrogation No. 477, Colonel Satake Kinji, 28 November 1945, https://dl.ndl.go.jp/info:ndljp/pid/4011397.
13. *Tokushu gijutsu kenkyu renryakusha no ken*, 4 September 1936, JACAR, Ref. C01004176400, JACAR; Kusaba, 'Tokushu,' 721; Ban, *Rikugun*, 122.

14. Ban, *Rikugun*, 114-17; Yamada, '*Kami*,' 104; Meiji, *Gaidobukku*, 16, 29; GHQ/TID, *Japanese*, 6-7, 12; Satake, 'Denpa', 573, 579-580; 'Okabe Kinjiro,' Kotobank, https://kotobank.jp/ (long URLs are abbreviated to save space)
15. Ban, *Rikugun*, 118-119; Shiina, 'Noborito,' 37; UEC Museum, 5.
16. Satake, 'Denpa,' 595-596; GHQ/TID, *Japanese*, 17-18; GHQ/AFPAC, *Report on Scientific Intelligence Survey in Japan, September and October 1945* (hereafter GHQ/AFPAC, *Report*), 1 November 1945, Volume II, Appendix, 1-E-2.
17. Shiina, 'Tenji,' 44. Sasaki Tadashi, later a vice president of Sharp Corporation, helped make the company a major postwar maker of microwave ovens, portable calculators, and other consumer electronics.
18. GHQ/TID, *Japanese*, 14, 18; GHQ/AFPAC, *Report*, Volume II, Appendix 1-E-1. Colonel Satake told Drs Griggs and Longacre of his work at Tama. *See* Colonel F.P. Munson, GHQ/SCAP G-2, to the Liaison Committee (Tokyo) for the Imperial Japanese Army, 6 November 1945, JACAR, Ref. C15010484000.
19. GHQ/TID, *Japanese*, 15, 18; GHQ/AFPAC, *Report*, Volume II, Appendix 1-E-3; Williams, 'Japan's,' 159.
20. Ban, *Rikugun*, 107-108.
21. Nagase-Reimer and Kawamura, 'Nihon,' 5-6.
22. Satake, 'Denpa,' 577.
23. Meiji, *Gaidobukku*, 16, 29.
24. Nagase-Reimer and Kawamura, 'Nihon,' 7. The IJA and IJN also pursued separate projects to develop an atom bomb. S*ee* the chapter 'Nuclear Energy and the Atomic Bomb' in Grunden, *Secret*. The lack of communication and cooperation between Japan's two military services, working separately to develop weapons rather than pool resources, is what Richard Samuels described in his history of the Japanese intelligence community as the problem of 'silos' or 'stovepipes.' *See* Samuels, *Special*, 16.
25. Ban, *Rikugun*, 123-24. Prince Takamatsu was a younger brother of Emperor Hirohito.
26. Lovell, *Of Spies*, 15; Grunden, *Secret*, 86, 106-107; Ban, *Rikugun*, 110; Yamada, *Kagaku*, 66.
27. Ban, *Rikugun*, 115, 122.
28. Satake, 'Denpa,' 595.

Chapter 3: Preparing Biological Weapons

1. Kim, 'I Found,' 33.
2. For poisoning in the Renaissance, *see* the chapters by Cathy Cobb and Sheila Barker in Wexler, *Toxicology*. On the pens, *see* Melton, *Ultimate Spy*, 191.
3. Houghton, *Nuking*, 21.
4. Grunden, *Secret*, 180, 182, 186-87; *Materials on the Trial of Former Servicemen of the Japanese Army Charged with Manufacturing and Employing Bacteriological Weapons* (hereafter, *Materials*), Moscow, Foreign Languages Publishing House, 1950, 4-15, 306.
5. Ban, *Rikugun*, 34.
6. Nagano/Akaho, *Kokosei*, 40, 99–100; Ban, *Rikugun* 77, 78, 81; Saito, *Boryakusen*, 134-35.
7. Tsukamoto, *Tenji*, 25-26.
8. Watanabe, 'Watakushi,' 96, 99. Meiji, *Gaidobukku*, 5.
9. Ban, *Rikugun,* 81-82.
10. Saito, *Boryakusen*, 126; Ban, *Rikugun,* 82; Tsuneishi and Asano, *Saikinsen*, 200-201. Sheldon H. Harris used material in Tsuneishi and Asano's book in his own, referring to Noborito as 'Kyu-Ken,' a Japanese abbreviation for the 9th ATRI. *See* Harris, *Factories*, 145-146.
11. 'Rikugun Noborito Kenkyujo (2): Chiratsuku "731 Butai" no kage,' *Tokyo Shimbun*, 12 August 1989.
12. Major Shishikura Juro, an officer of the Military Affairs Section of the Military Affairs Bureau, told a writer after the war that Second Section was developing weapons to destroy sorghum, a staple crop in northern China. *See* Saito, *Supai*, 174. Noborito technician Matsukawa Hitoshi claimed that corn, potato, and wheat, major crops in the United States, were focal points of research in Second Section's Group 6 in late 1941. *See* Ban, *Rikugun*, 87, and Yamada and Meiji Daigaku, *Rikugun*, 119-120.
13. Nagano/Akaho, *Kokosei*, 80.
14. *Materials*, 22, 57-58; Mizutani, '*1644*,' 51. The novelist Yoshimura Akira recounted this episode in his 1975 novel of IJA biological warfare, *Nomi to bakudan*; Mercado, *Shadow*, 133.
15. Major General Imai Takeo wrote of IJA peace operations in his 1967 history *Showa no boryaku.*
16. Okada, 'Shina,' 220-223.
17. 'Kyu Rikugun "Sho Kaiseki dokusatsu" o keikaku,' *Tokyo Shimbun*, 31 August 1998.

18. Martin, 'Shield,' 101-102, 131, 134-35.
19. Yick, 'Communist-Puppet,' 85, note 3.
20. Ban, *Rikugun*, 95–97.
21. The transcripts of the Khabarovsk trials refer at numerous points to Unit 100 activities, including human experimentation. *See* also Harris, *Factories*, 119, 122, 124-126, 128-131; Ban, *Rikugun,* 96-97.
22. Ban, *Rikugun*, 27; Yamada and Meiji Daigaku, *Rikugun*, 136; Kawashima, *Ushi*.
23. Kuba, *Rikugun,* 2; Ban, *Rikugun*, 163. My thanks to Ross Coen and Amanda McVety for sending me Kuba's document.
24. Kuba, *Rikugun,* 2; Nihon Tosho Centre, *Kyu*, 531; Nakamura, *Ichi*, 14; Tsuchiya, 'Dobutsu,' 11. In 1910, the Empire of Japan annexed Korea, known widely thereafter in English as Chosen, the Japanese pronunciation of the Korean word for Korea. With Japanese the empire's official language, Korean place names appeared in English in Japanese pronunciation. Chosen's capital was Keijo (Kyongsong, later Seoul). Other prominent cities were Heijo (Pyongyang), Fusan (Pusan), and Genzan (Wonsan). Korea was liberated in 1945, but Japanese names returned several years later on maps sent from Japan to United Nations forces in the Korean War. Under the influence of the Japanese maps, government and media reports described US Marines fighting a hard-pressed withdrawal from the Chosin (Changjin) Reservoir (hence the references to the 'Frozen Chosin') and the 1950 evacuation of UN forces and Korean refugees from the port of Konan (Hungnam).
25. Nihon Tosho Centre, *Kyu*, 531; Isayama Isaburo, https://www.jacar.go.jp/; Nakamura, *Ichi*, 215; Otake, 'Nihon,' 114-115.
26. Kuba wrote in his document of a Professor Naito Motoo of Tokyo Imperial University as the Soviet expert. Ban, who incorporated most of Kuba's document into his book, changed Naito to Sasaki. Kuba, *Rikugun*, 2-4; Ban, *Rikugun*, 99-100. A 1938 paper that Hotta co-authored indicated that he worked in the Vaccine Department. Akazawa Sasao and Hotta Tokuro, 'Hito todoku o motteseru dobutsu jikken toku ni sono gyutoka ni tsuite,' *Nihon Juigaku Zasshi*, 1938, 241-255.
27. Tsuchiya, 'Dobutsu,' 12; Nakamura, Ichi, 3, 13, 215; Ban, *Rikugun*, 99; McVety, *Rinderpest*, 31.
28. Nomura, 'Zaidan,' 67; Kim, 'Toi,' 204; Im, *Gendai*, 65; Yi, *Pukhan*, 219. Kim's article first appeared in Korean in the December 1993 issue of *Choguk*, a monthly magazine affiliated with the pro-

Pyongyang General Association of Korean Residents in Japan (known in Japanese as the Chosen Soren).

29. Nomura, 'Zaidan,' 67; Kim, 'I Found,' 33.
30. Kuba, *Rikugun*, 4, 11–13.

Chapter 4: Bombing America by Balloon

1. Kusaba, 'Fusen,' 72.
2. Sato, *Dai*, 283. A lawyer asked Sato at the postwar International Military Tribunal for the Far East if he had called for attacking New York's skyscrapers and, if so, where he had done so. Sato replied that he had done so in the room where he was on trial. Sato wrote of the situation as 'an irony of destiny.'
3. Kusaba, 'Fusen,' 73; Kusaba, 'Beihondo,' 526-529.
4. Kusaba, 'Beihondo,' 529, 531; Yamamoto, 'Nisesatsu,' 105; Yamamoto, *Rikugun*, 190; Kinoshita, *Kesareta*, 138; Nagano/Akaho, *Kokosei*, 76.
5. Goto, 'Genbaku,' 182.
6. *A Report on Japanese Free Balloons*, Joint Army-Navy Release (hereafter *Free Balloons*), 8 February 1946. For balloon mechanical details and photographs, *see* Mikesh, *Japan's*. The writer Saito Michinori put the number of sandbags at 32. *See* Saito, *Boryakusen*, 61.
7. According to Dr LaPaz, Japanese used wet-cell batteries, rather than dry ones; some batteries froze in flight, preventing self-immolation. *See* LaPaz, 'Now,' 10.
8. Kusaba, 'Fusen,' 72.
9. Goto, 'Genbaku,' 183; Maema, *Fugaku*, Vol. 2, 303.
10. Watanabe and Yamada, 'Moto fusen,' 89.
11. Suzuki, *Fusen*, 127-129, 134; Sato, *Dai*, 284; Kusaba, 'Beihondo,' 542; Maema, *Fugaku*, Vol. 2, 309.
12. Cook, *Japan*, 188-191; Suzuki, *Fusen*, 128-129, 133-134; Kusuyama, '*Rikugun*,' 100.
13. Sato, *Dai*, 60-61, 284; Kinoshita, *Kesareta*, 124–126. On the same day as General Umezu's order, Vice Admiral Onishi Takijiro, commander of the First Air Fleet in the Philippines, ordered for the first time IJN pilots to fly into American warships in what became known as *kamikaze* (divine wind) attacks. The IJA and IJN, unable to counter the US counteroffensive by conventional means, resorted that day to unconventional ones. See Maema, *Fugaku*, Vol. 2, 289.

14. Nagano/Akaho, *Kokosei*, 77; Sato, *Dai*, 284.
15. LaPaz, 'Now,' 9-10.
16. Mikesh, *Japan's*, 25; 'Subj: Balloon and Appurtenances, Japanese, from Kailua, Hawaii,' Naval Research Laboratory, 15 January 1945, RG 499, box 41, NACP; 'Japanese Aircraft Makers Plates and Markings, Japanese Free Balloons,' Military Intelligence Report, No. 78, 16 May 1945, RG 319, box 7353, NACP; 'Subject: Capability of Japanese to Attack US by Airborne Devices,' Memorandum for the Chief, Military Intelligence Service, from G-2 Military Intelligence Division, War Department General Staff, 21 December 1944, RG 319, box 7350, NACP; 'Subject: Japanese Balloons,' Memorandum from Clayton Bissell, Major General, AC of S, G-2, for Director of Naval Intelligence; Assistant Chief of Air Staff, Intelligence, 24 December 1944, RG 319, box 7350, NACP. According to Bissell's memorandum, the recovered bomb fragments were shipped to Aberdeen Proving Ground in Maryland for further examination.
17. 'Subj: Balloon and Appurtenances, Japanese, from Kailua, Hawaii,' Naval Research Laboratory, 15 January 1945, RG 499, box 41, NACP; LaPaz, 'Now,' 10-11.
18. 'Subject: Capability of Japanese to Attack US by Airborne Devices,' Memorandum for the Chief, Military Intelligence Service, from G-2, Military Intelligence Division, War Department General Staff, 26 December 1944, RG 319, box 7350, NACP.
19. 'Subject: Japanese Balloons,' Memorandum No. 10, Military Intelligence Service, War Department, 13 January 1945. RG 319, box 7350, NACP.
20. 'Tokyo Radio,' 17 February 1945, RG 499, box 37, NACP. The unattributed radio broadcast summary likely came from the Foreign Broadcast Intelligence Service (FBIS), which monitored Axis and other foreign broadcasts. Tokyo learnt of the Montana incident, reported in US media, via the Chinese newspaper *Takungpao*. See Mikesh, *Japan's*, 25, 37. On IJA elation over balloons reaching the United States, *see* Kusaba, 'Fusen,' 76.
21. 'A Report on the Office of Censorship,' Washington, DC, Government Printing Office, 1945, 40; Coen, *Fu-go,* 149; Kusaba, 'Fusen,' 76.
22. Mikesh, *Japan's*, 29; 'Subject: Possible Introduction of Disease-Producing Agents by Japanese Balloons,' Adjutant General's Office, War Department, 12 February 1945, RG319, box 41, NACP.

23. Kusaba, 'Beihondo,' 540.
24. 'Subject: Power Outage 10 March, 1945,' from Colonel F.T. Matthias, Corps of Engineers, Hanford Engineer Works, US Engineer Office, War Department, to Major General L.R. Groves, P.O. Box 2610, Washington, DC, 29 March 1945, RG 77, NACP. Burt Pierard, historian of the B Reactor Museum Association (BRMA), disputes the memo's description of the balloon contact with the wires resulting in a line surge, writing that the contact caused a dip 'obviously severe enough to trip the Power Failure to the Process Pump Building Protective Relays at each of the reactors (the only electrical disturbance devices that could cause a full SCRAM). The simultaneous trip of all three Reactors is further indication of reaction to the system wide "dip."'*See* Pierard, *Moderator*, 3.
25. "The Atomic Bombing of Nagasaki," www.osti.gov.
26. Pierard, *Moderator*, 4.
27. 'Japanese Bomb-Laden Balloons Proved Fizzle as War Weapon', *New York Times*, 16 August 1945.

Chapter 5: Counterfeiting Currencies, Forging Documents

1. Yamamoto, *Rikugun*, 11.
2. Mendez, *Master*, 44. In 2007, the Austrian film *Die Fälscher* (The Counterfeiters) told the story of Germany's secret plan, Operation Bernhard, to destabilise the British economy with counterfeit pounds.
3. Kinoshita, *Kesareta*, 148-149; US Department of Commerce, 'China's Currency Complexities: Basic Facts and Current Trends,' *Foreign Commerce Weekly*, 5 July 1941, 5-7, 19.
4. Sun, *Art*, 74.
5. Yamamoto, *Rikugun*, 174; Fang, 'Senchu,' 86.
6. Yamamoto, *Rikugun*, 56-63, 245.
7. US Department of State. *Foreign Relations of the United States Diplomatic Papers, 1941, The Far East*, Volume IV, Documents 34 and 35, https://history.state.gov; Iwakuro, *Shingaporu*, 207.
8. Yamamoto, *Rikugun*, 63-64; Iwakuro, *Iwakuro*, 139. According to Okada, Iwakuro, a staff officer of the Kwantung Army, had employed Sakata soon after the Manchurian Incident in an unspecified operation. *See* Yamamoto, 'Nisesatsu,' 106.

9. Okada, 'Chugoku,' 42; Okada Yoshimasa, 'Shina,' 220-222. Iwakuro had discussed with Sakata in Tokyo in late 1937 killing Chiang and counterfeiting Chinese bank notes. *See* Kumano, '*Sakata*,' 49-53.
10. Yamamoto, 'Nisesatsu,' 103; Oshima, 'Insatsu,' article 1, 31-32; Oshima, 'Insatsu,' article 2, 28; Shiina, 'Noborito,' 46.
11. Yamamoto, 'Nisesatsu,' 104; Nakano Koyukai, *Rikugun*, 338. Sakata, was working under the Chinese name Tian Cheng from the Hong Kong Hotel with local triads to organise fifth-column activities. Jailed by the British authorities, he escaped and slipped back into China. *See* Philip Snow, *The Fall of Hong Kong*, 37-38.
12. Yamamoto, *Rikugun*, 84-85; Iwakuro, *Iwakuro*, 136.
13. Oshima, 'Insatsu,' article 1, 34; Iwakuro, *Iwakuro*, 135; Okada, 'Chugoku,' 49-50.
14. Kuba, *Rikugun*, 2.
15. Yamamoto, *Rikugun*, 92, 98; Ban, *Rikugun*, 34; Fujiwara, *Ryukonroku*, 109-110; Kusuyama, '*Rikugun*,' 139.
16. Yamamoto, *Rikugun*, 109-110; Imai, *Showa*, 195.
17. Nakano Koyukai, *Rikugun*, 133. According to Yamamoto Kenzo, couriers also sailed from the Japanese ports of Moji and Shimonoseki. *See* Yamamoto, *Rikugun*, 123-124.
18. Saito, *Shogen*, 173-174; Oshima, 'Insatsu,' article 2, 28. Tian, as noted earlier, was an alias for Sakata.
19. Yamamoto, *Rikugun*, 128-129.
20. According to Yamamoto, Sakata operated from Tu's mansion and a building in the French Concession that housed the Hua Shin and Min Hua companies. *See* Yamamoto, *Rikugun*, 126-128; Okada, 'Chugoku,' 49. Okada remarked after the war that the Matsu Agency had contacts with Chungking and even took orders for such items as lipstick and dress material for the wives of Nationalist leaders in Chungking. Okada, 'Shina,' 229.
21. Saito, *Shogen*, 176. Perhaps it was no coincidence that Okada received orders in June 1943 to report to Canton as a senior staff officer of the IJA Twenty-Third Army; Kumano, '*Sakata*,' 46-47.
22. Yamamoto, *Rikugun*, 130–131. In return, Hsieh Wen-ta, a division commander in the Nanking regime's military, provided from his position in Chekiang Province protection for the transport of tungsten and other needed materials procured in Operation Sugi. *See* Okada, 'Shina,' 229.

23. Saito, *Shogen*, 174; Yamamoto, 'Nisesatsu,' 104-105.
24. Toye, *Springing*, xiii; Nakano Koyukai, *Rikugun*, 401.
25. Mercado, *Shadow*, 66-67.
26. Yamamoto, 'Nisesatsu,' 104; Saito, *Shogen*, 176-177; Nakano Koyukai, *Rikugun*, 136; Iwakuro, *Iwakuro*, 135, 217; *The Japanese Intelligence System*, MIS/WSGS, 4 September 1945, 38, RG 457, box 90, NACP.
27. Yamamoto, *Rikugun*, 189; Yamamoto, 'Nisesatsu,' 104-105.
28. Iwakuro, *Iwakuro*, 136; Nakano Koyukai, *Rikugun*, 136.
29. Lovell, *Of Spies*, 23.
30. Yamamoto, *Rikugun*, 190.
31. Kusuyama, '*Rikugun*,' 154; Yamamoto, 'Nisesatsu,' 104.

Chapter 6: Withdrawing for Battle, Destroying Evidence

1. Ota, *731*, 6-7, 189. Ota, interviewing Niizuma, learnt that Niizuma wrote the memo to destroy incriminating evidence of the IJA biological weapon programmes. *See* also Kinoshita, 'Nagano-ken,' 3.
2. Mercado, *Shadow*, 132-134. Lieutenant Onoda Hiroo of the Nakano School was famous after the war for keeping to his wartime mission on the Philippine island of Lubang until his former commanding officer relieved him of his duty in 1974. *See* Onoda's autobiography, *No Surrender*.
3. AFPAC, 'Japanese Wartime Military Electronics and Communications, Section II, Japanese Military Research Laboratories and Projects,' 1 April 1946, 35, RG 319, box 2121, NACP.
4. Hatakeyama, *Rikugun*, 191; Yoshinaga, 'Nihon', 18; Nagano/Akaho, *Kokosei*, 106.
5. Yoshida, 'Pasokon,' 137-138.
6. Saito, *Boryakusen*, 94.
7. Mercado, *Shadow*, 125-126; Matsushiro Imperial Underground Headquarters, https://matsushiro.org/en.
8. Oshima, 'Insatsu,' article 1, 32; Ban, *Rikugun*, 192; GHQ/TID, *Japanese,* 16.
9. AOAH. *Rikugun gijutsu kenkyujo no bu/fuhyo: Dai Kyu Rikugun Gijutsu Kenkyujo jininhyo*, 31 August 1945, JACAR, Ref. C15010410200.

10. Kinoshita, *Kesareta*, 310; Nagano/Akaho, *Kokosei*, 107, 120.
11. Satake, 'Denpa,' 577.
12. Kinoshita, *Kesareta*, 356; Nagano/Akaho, *Kokosei*, 59; GHQ/TID, *Japanese*, 12, 16, 19; Satake, 'Denpa,' 598.
13. NAVTECHJAP, *Japanese Electronic Tubes* November 1945 (hereafter NAVTECHJAP, *Japanese*), 27, www.fischer-tropsch.org; GHQ/TID, *Japanese Research*, 19.
14. GHQ/AFPAC, *Report*, Volume I, 6, 42; Shiina, 'Noborito,' 44.
15. Yonehara Norihiko, 'Rikugun Nakano Gakko no kyokasho 8 shu hakken,' *Asahi Shimbun*, 7 February 2005.
16. Nakamura, *Ichi*, 49; Grant and Guelich, 'Air,' 25. Dr Nakamura did not give a date for the departure for the interior, but the mining of Fusan took place in Phase V, 9 July to 15 August, of Operation Starvation. *See* AAF, *Impact: Air Victory Over Japan*, Volume 3, No. 9 September-October 1945, 50.
17. Kuba, *Rikugun*, 13. Mercado, *Shadow*, 133; Kusuyama, '*Rikugun*,'193-195; Hori, *Daihonei*, 223-224. AGS Second Bureau estimated that the Soviet Union would attack Japan in November, according to the bureau's Lieutenant Colonel Asai Isamu. *See* USSBS, 'Subject: Organization and Operation of TOKUMU KIKAN in Manchuria,' 372–5 https://dl.ndl.go.jp/pid/4012618/1/1
18. Kusuyama, '*Rikugun*,' 109-111.
19. Watanabe and Yamada, 'Moto fusen,' 94-95, 101-102, 124; Tsukamoto, 'Fusen,' 34, 36.
20. *Materials*, 9, 43, 47.
21. 'Report to V. Bush on the Japanese Balloon Problem by an OSRD Special Committee,' 1, RG 165, box 318, NACP.
22. 'Subject: Capability of Japanese to Attack US by Airborne Devices,' 26 December 1944, RG 391, box 7350, NACP; 'Subject: Japanese Balloons,' 13 January 1945, RG 319, box 7350, NACP; 'Subject: Possible Introduction of Disease-Producing Agents by Japanese Balloons,' 12 February 1945, RG 319, box 41, NACP.
23. Subject: Indoctrination of Selected Personnel in Aspects of BW of Japanese Balloons,' 15 February 1945, RG 319, Box 41, NACP; LaPaz, 'Now,' 11; Brophy, *Chemical*, 115.
24. *Free Balloons*, 8 February 1946; LaPaz, 'Now,' 9, 11.
25. 'Air Information,' 12 May 1945, RG 319, box 2097, NACP; 'Subject: Release of Photographs of Japanese Free Balloons,' 25 May 1945, RG 499, box 40, NACP.

26. LaPaz, 'Now,' 11; Mikesh, *Japan's*, 29; Nagano/Akaho, *Kokosei*, 75; *Joint Western Sea Frontier – Western Defense Command Plan Covering Defences Against Japanese Free Balloons, Short title 'BD-1,'* 15 August 1945, Annex F, 1, RG 165, box 1877, NACP.
27. NAVTECHJAP, *Japanese*, Enclosure D, 27, www.fischer-tropsch.org.
28. Kuba, *Rikugun*, 12–13; 'Rikugun Noborito Kenkyujo (3): Sokai saki de mo saikin heiki kaihatsu,' *Tokyo Shimbun*, 13 August 1989.
29. Saito, *Boryakusen*, 97.
30. Nagano/Akaho, *Kokosei*, 116.
31. Kinoshita, *Kesareta*, 336-338.
32. Oshima, 'Insatsu,' article 2, 30; Yamamoto, *Rikugun*, 108, 118, 167.
33. Ota, *731*, 53-54, 189.
34. Kinoshita, 'Nagano-ken,' 3-4; Ota, *731*, 189.

Chapter 7: Dealing With the Victors in Occupied Japan and Korea

1. Arisue, *Arisue Kikan*, 66-67.
2. Nouzille, Huwart, 'Comment,' 53–54; Faligot, Krop, *Piscine*, 47.
3. Captain Donald B. Sumners to Lieutenant Colonel John M. Jeffries, Subject: Final Summary Report 27 September 1945, RG 226, box 21, NACP; Brophy, *Chemical*, 114-115.
4. NARA Staff, 'Japanese War Crimes Records at the National Archives: Research Starting Points,' in Drea, *Researching*, 98-99; *Materials*.
5. Gehlen, *Service* 8–9; Mercado, *Shadow*, 211-214, 218-220; Torii, *Nihon*, 58-59; Kotani, *Japanese*, 26.
6. 'Harubin tokumu kikan,' 45; Kato, *Rikugun*, 181.
7. Hata, *Nihon*, 10; Arisue, *Arisue Kikan*, 57-58, 66-67. Kawabe, *Kawabe*, 184; Mashbir, *I Was*, 296; Hori Eizo, an intelligence officer in Arisue's Second Bureau, recalled that the director, ardently pro-Axis, had pushed for Japan to join Germany and Italy in alliance. His passion for Mussolini's Italy was such that other IJA officers are said to have wondered aloud in sarcasm whether he was Italian. *See* Hori, *Daihonei*, 78.
8. Arisue, *Arisue Seizo*, 409-411; *Arisue Kikan*, 164–165.
9. SCAP, 'Application for Remaining in Office' 20 April 1947, RG 331, box 2275V, NACP; Hata, *Nihon*, 10.
10. Kawabe, *Kawabe*, 195; Willoughby, *Maneuver,* 196; Arisue, *Seiji*, 264.

11. Arisue, *Arisue Kikan*, 249.
12. Ota, *731*, 67, 68, 71, 132-134. The deception lasted longer on BW activities against the United States. Colonel Yamada Sakura in April 1948, requested the discretion of Japanese police investigating a poisoning case, explaining that IJA officers had kept from the Americans intelligence on Groups 6 and 7 of Noborito's Second Section, which had worked on biological weapons to attack the United States, and on Third Section. *See* Tsukamoto, '"*Kai*,"' 20; Captain Hartwig Kuhlenbeck to Lieutenant Colonel G.W. Anderson, 12 October 1945, in Cunliffe, *Select*, 24.
13. ATIS, *Alphabetical List of Japanese Army Officers*, May 1943. Copy on file at CMH. War Department. 'Military Research Bulletin No. 21: Japanese Intelligence,' 15 August 1945. On file at CMH. By late September 1945, an appendix to a military report described Noborito's purpose as 'radio research.' *See* Technical Intelligence Headquarters, XI Corps, *Preliminary Report on Japanese Army Experimental Stations*, 26 September 1945, RG 496, box 339, NACP.
14. A 1945 American-British document listed Shinoda and the directors of Noborito First and Second Sections but omitted Third Section. *See* War Department, *List of Staff of the Japanese Army*, 1 August 1945, https://dl.ndl.go.jp/pid/4009938/1/1. GHQ/AFPAC, *Report*, Volume I, 10, RG 165, box 2055, NACP.
15. AFPAC. *Basic Outline Plan for 'Blacklist' Operations,* Annex 5d, 'Basic Intelligence Plan', 8 August 1945.
16. Willoughby memo to Sutherland, 'Subject: Punitive Features of Annexes to Operations Instructions,' 27 August 1945. Copy on file at CMH.
17. 'BCOF Faced Many Difficulties In Its Six Months of Occupation,' *BCON*, 31 August 1946, p.5; US Army, *Reports of General MacArthur, (I) Supplement, MacArthur in Japan: The Occupation: The Military Phase Volume I,* Washington, DC, Government Printing Office, 1966, 134, 138. On Nakano School plans in the event of a punitive occupation, *see* Mercado, *Shadow*, 175-177.
18. Dr Karl Compton, initial survey team leader, was in Japan for the first two weeks (until 24 September). The several members of the technical staff were civilians and military officers from CWS, OSRD, and other military organisations.
19. AFPAC, *Report*, Volume I, 9, 27, 42, 43, Volume II, I-D-42, RG 165, box 2055, NACP.

20. Saito, *Boryakusen*, 96-97, 99. Kojima Tatsuji, deputy director of Noborito's factory in Nakazawa, Nagano Prefecture, recalled a CIC team coming to Nakazawa by jeep around 10 September. He told Saito of his surprise on how much intelligence the Americans already possessed so early in the Occupation. *See* also 'Rikugun Noborito Kenkyujo (1): Nihon ni mo "minna goroshi heiki,"' *Tokyo Shimbun*, 11 August 1989.
21. Saito, *Boryakusen*, 136; Kinoshita, *Kesareta*, 358.
22. Ban, *Rikugun*, 196. Ban identified the unit, apparently in error, as the 411th CIC Detachment.
23. 'O C ORD O ADV to G-2/WDIT,' 10 December 1945; 'Fukurenho Dai 596-go', 11 December 1945; 'O C Sig O/TLID to G-2/WDIT,' 7 January 1946. JACAR, Refs. C15010484100, C15010162300, C15010484200, and C15010484300.
24. Colonel Creswell to War Department G-2, 'Subject: Japanese Use of Secret Ink', 21 February 1946, RG 331, box 8517, NACP.
25. CCD, *Interrogation of the Two Foremost Japanese Secret Writing Chemists*, 10 July 1946, 3, 4, 6, RG 331, box 8554, NACP. My thanks to Tsukamoto Yuriko for the archival information that pointed me to the location of the interrogation report. Saito, *Boryakusen*, 139–140.
26. NAVTECHJAP, *Japanese Use of Balloons for Barrage, Bombing and Aerology*, December 1945, 2. The report was produced in line with a target list from 4 September 1945. www.fischer-tropsch.org.
27. Kusaba, 'Beihondo,' 532; AFPAC, *Report*, interview of Dr Arakawa and Lieutenant Colonel Niizuma by Major Hewlett, Dr Griggs, Tec 4 Yagi, Volume III, Appendix 23-1 – 23-2, 13 October 1945, RG 165, box 2055, NACP; GHQ/SCAP, George E. Weidner, 'Subject: Technological Survey; Ten Metre Japanese Paper Bombing Balloons,' 28 November 1945, RG 496, box 341, NACP. For several meetings on Noborito and biological and chemical warfare *see*, Arisue Agency reports, numbers 264 (24 October 1945), 340 (8 November 1945), and 383 (15 November 1945). JACAR, Refs. C15010238500, C15010246200, and C15010250100.
28. AFPAC, *Report*, interview of Lieutenant Colonel Kunitake and Major Inoue by Lieutenant Colonel Sanders, Major Skipper, and Tec 4 Yagi,' 19 September 1945, Volume II, Appendix 1-B-1, RG 165, box 2055, NACP. The fifth, and final, volume of the report, on biological warfare, included interviews with personnel from Unit 731, but none from Noborito.

29. Saito, *Boryakusen*, 63.
30. Goto, 'Genbaku,' 183. The deception that the Japanese Army never intended to use the bombing balloons for biological warfare continued for years. Kusaba repeated the script in the late 1960s to an American NCO, who, repeated it in a professional journal of the US Air Force. *See* Cornelius Conley, 'The Great Japanese Balloon Offensive,' *Air University Review*, January-February 1968, 68-83. Indeed, IJA veterans for years denied there had been an offensive BW programme. *See* Iwakuro, 'Junbi,' 21, in which even as late as 1956 Iwakuro denied that the Japanese Army had conducted offensive BW research and development.
31. NAVTECHJAP, *Japanese*, November 1945, 3-4, 7, www.fischer-tropsch.org; GHQ/TID, *Japanese*, 24 June 1949, RG 319, box 2119, NACP; 2d Lieutenant Harry B. Bekkar to Director, TLID, 'Subject: Investigation of Communication Research and Production at Nakazawa Factory Branch of the 9th Military Technical Laboratory, Nagano Prefecture, and the Nippon Koshuha Co. Ikeda, Nagano Prefecture,' 26 January 1946, RG 496, box 342, NACP.
32. AFPAC, *Report*, interview of Lieutenant General Shinoda, Major General Kusaka, and Lieutenant Colonel Niizuma by Drs Griggs, Moreland, and Stephenson, Volume II, Appendix 1-E-1–1-E-2, 13 October 1945, RG 165, box 2055, NACP.
33. Nomura, 'Zaidan,' 67; Otake, 'Nihon,' 117; Nakamura, *Ichi*, 215. The former OGGC Livestock Hygiene Research Institute appears as the 'Fusan Laboratory' in AFPAC, *Summation No. 6: United States Army Military Government Activities in Korea*, March 1946, 21. https://digirepo.nlm.nih.gov; Nakamura, *Ichi*, 51. US authorities ordered two of their Japanese assistants also to stay behind.
34. Nomura, 'Zaidan,' 67. Nakamura recalled US authorities threatening anyone caught returning to Korea from Japan with deportation to Manchuria, then under Soviet occupation. *See* Nakamura, *Ichi*, 51. Nakamura wrote that Korean staff of the Livestock Hygiene Research Institute were kind and understanding, allowing Nakamura and other Japanese staff to travel between Japan and Korea. *See* Nakamura, *Ichi*, 52.
35. Kim, 'Toi,' 204-06; Ri, 'Han,' 25; Yi, *Pukhan*, 220. Kim did not mention whether or not Americans took any of Kim's subordinates to the United States, but such an action would have been consistent with Washington's postwar debriefing and employment of German

and Japanese scientists and technicians in the United States. In Yi's account, Kim rejected a US pitch to work at a 'military microbial research institute,' presumably in the United States.

Chapter 8: Working With the Americans

1. Saito, *Boryakusen*, 109.
2. Arisue, *Arisue Kikan*, 85.
3. Takai, 'Saitenken!', 203; Thomas R. Johnson, *American Cryptology during the Cold War, 1945-1989, Book I: The Struggle for Centralization, 1945-1960*, NSA, 1995, 49, www.governmentattic.org.
4. Petersen, 'Intelligence,' in Drea, *Researching*, 199-206.
5. Kinoshita Kenzo, author of an early history of Noborito, refers to the document as revealing "only harmless and inoffensive research", in Kinoshita, 'Nagano-ken', 5–6.
6. The British must have known that the Japanese were counterfeiting rupees, as they had caught some number of IJA agents in India. *See* Howard, *British*, 207-209.
7. Saito, *Boryakusen*, 100, 193. The US Army also enlisted Major Ban in 1948 in postwar clandestine work through 'give and take,' reportedly after he had admitted to Japanese police his wartime involvement in human experimentation in China and, presumably, GHQ/SCAP had learnt of it. Speculation in Japan is that Ban in April 1948 told police of his involvement in human experiments; the police then would have relayed what he said to SCAP, who then sent representatives to Ban to discuss 'give and take.' *See* Yamada and Meiji, 'Noborito,' 118; Kusuyama, '*Rikugun*,' 168.
8. Saito, *Boryakusen*, 108, 110, 113, 118-119; Ban, *Rikugun*, 147; Kinoshita, *Kesareta*, 385. GPSO also stood for Government Printing and Supply Office, according to a labour contract that Kinoshita Chokichi, a veteran of Noborito's Third Section, signed in 1955. Unclear is whether Kinoshita's GPSO, located not at Yokosuka but at North Camp Drake, a US military base in Saitama Prefecture, was a separate organisation from the Yokosuka GPSO. See Tsukamoto, 'Yokosuka,' 131.
9. Yi, *KLO*, 147. Documents came by various routes from Korea to the GPSO, according to Yi. One was the KLO, which operated under Major General Willoughby. KLO operatives would collect

documents from the enemy – captured, killed or otherwise dead – then send the documents to 'Tokyo,' almost certainly the GPSO, where document specialists would forge identical documents for KLO and other operatives to use in missions behind enemy lines. Kinoshita, *Kesareta*, 388. Kinoshita writes, without sourcing, that the Canon Agency, 308th CIC Detachment, CCRAK's 8240 AU, and CCRAFE's AU 8177 received GPSO documents.

10. Chong, *Chong*, 247. The 527th and 589th QMTIDs managed the uniforms taken from Chinese and Korean prisoners of war. 'CCF and NKA' uniforms referred to the those of the CPV and KPA. The uniforms went as a 'first priority' to fulfil 'local clandestine requirements.' *See* Colonel Chas. M. Myrick, G-2, to Office of the Quartermaster General, 'Subject: Request for CCF and NKA Uniforms,' 8 August 1952, https://dl.ndl.go.jp/pid/8341727/1/6. It is unclear whether or not the two QMTIDs also managed the forged documents that went with the captured uniforms to meet 'clandestine requirements.'
11. Clark, *Secrets*, 8, 17; Yon, *Kyanon*, 151; Kye, *Maegado*, 162.
12. Willoughby, *GHQ*, 285; Saito, *Boryakusen*, 118.
13. Yi, *KLO*, 183.
14. Yon, *Kyanon*, 199-213.
15. Saito, *Boryakusen*, 109, 113, 165.
16. In the Korean War, for example, a half dozen Nisei veterans of the Japanese Army worked at the United Nations prisoner interrogation facility at Tongnae alongside 100 or so *Kibei Nisei*. The latter Nisei were those who, after residing in Japan, had returned to the United States before Pearl Harbor. Unlike Nisei resident in Japan, *Kibei* Nisei served in the Second World War on the American side. *See* Yanagida, *Nisei*, 222-223.
17. Takai, 'Saitenken!', 208; Seino Tatsuo, who later became president of what has been known since 1983 as PASCO Corporation, established in 1953 as Pacific Airborne Surveying Co., Ltd., was an IJA surveyor from ILS who later worked for the US Army 64th Engineer Battalion. Yamamoto Yoshio, a surveyor for Imperial Japanese Airways, joined a number of his colleagues after the war at Isetan Department Store, working for the 64th. *See* www.jsokuryou.jp. ROK Army General Paek Son-yop recalled his first encounter of the war with maps showing Korean place names in English and Japanese, but not in Korean. *See* Paek Son-yop. *From*, 50.

18. Yamamoto, *Kenetsukan*, 11.
19. Okubo, *Uminari no hibi*, 299.
20. Saito, *Boryakusen*, 114.
21. One expert estimated that approximately 30 Japanese were working for the GPSO circa 1952. *See* Yamada Akira, '8 gatsu,' 136. Ban, *Rikugun*, 203; Tsukamoto, 'Moto joinra,' 84-85.
22. Tsukamoto, 'Moto joinra,' 84, 99 note 11; Ban, *Rikugun*, 203.
23. Iwakuro, *Iwakuro*, 135.
24. Saito, *Boryakusen*, 115.
25. Tsukamoto, 'Moto joinra,' 95; Friedman, 'Propaganda.' Moscow radio reported in November 1966 'a new and unseemly action by the US aggressors in Vietnam,' that of infiltrating counterfeit North Vietnamese bank notes into the DRV with the intent 'to cause inflation in the country so as to disrupt economic life.' *See* 'Currency Scheme' in FBIS, *Daily Report: Foreign Radio Broadcasts*, No. 226 – 1966, 22 November 1966, BB21, via https://books.google.com.
26. Saito, *Boryakusen*, 115, 166; GIA, 'Focus On: Kenzo Yamamoto,' *In Focus*, July 1984, 10. My thanks to the GIA for sending this publication.
27. Iwakuro, *Iwakuro*, 140; GIA, 'Focus on Kenzo Yamamoto,' *In Focus*, July 1984, 9.

Chapter 9: Rebuilding Japan

1. Oshima, 'Insatsu,' article 2, 30.
2. Hata, *Nihon*, 10; Arisue, *Arisue Kikan*, 254-255; Asahi Shinbunsha, ed. *Gendai*, 62; 'Goyu Renmei no enkaku,' www.goyuren.jp. Arisue wrote at length of his involvement in postwar veteran organisations, as well as of his connections to veteran movements in the ROK and the ROC. See Arisue, *Seiji*, 402-417.
3. Asahi Shinbunsha, ed., *Gendai*, 157. On Iwakuro working in the shadows with the Americans and the twenty-year gap between the end of his military career and his founding of Kyoto Sangyo University, *see* Kawai, 'Ichi.'31-37.
4. Yamamoto, 'Nisesatsu,' 103.
5. Web page of the Society of Fibre Science and Technology, Japan, www.fiber.or.jp/jpn/overview/chair.html.
6. Kojunsha, ed., *Nihon*, 1978, 138.

7. GHQ/TID, *Japanese*, II.
8. Kojunsha, ed., *Nihon*, 1971, 96.
9. Goto, 'Genbaku,' 182; Ota, *731*, 15, 18, 44. Niizuma, after overseeing IJA wartime special research, worked after the war as a technical official in the DA Central Technical Institute, researching nuclear weapons and guided missiles in the National Defence College. He rose to the rank of senior researcher in the First Research Centre of the DA Technical Research and Development Institute before retiring in 1970.
10. Miyamoto, 'Enka,' 58; Kojunsha, ed. *Nihon,* 1992, 96.
11. Ban, *Rikugun*, 203; Hashizume Kiyoshi, 'Studies on Organic Pigments of Quinacridone Series. I,' 20 January 1961, www.jstage.jst.go.jp; Saito, *Boryakusen*, 139.
12. Saito, *Boryakusen*, 78.
13. Kuba, *Rikugun*, 14. His document appeared in a shorter version that omitted the original reference to the 'annihilation' of Imperial Japan's enemies. *See* 'Kuba Noboru no shuki' in Ban, *Rikugun*, 97-105.
14. Oshima, 'Insatsu,' article 2, 30.
15. GIA, 'Focus On: Kenzo Yamamoto,' *In Focus*, July 1984, 9-10. AGT changed its name in 1993 to GIA Japan.
16. Bai Juyi, 'Bidding Farewell to Weizhi [Yuan Zhen] at the Lishui River on the 30th Day of the Third month of the Tenth Year,' in Yamamoto, *Rikugun*, 1.
17. Kojunsha, *Nihon*, 1980, I-137; Oshima, 'Insatsu,' article 2, 30; Oshima, 'Insatsu,' article 1, 33–34. Oshima also recalled with pride the day in the war that Field Marshal Sugiyama Hajime, on a tour of Noborito, praised Oshima for his work.
18. Oshima Yasuhiro, 'Kyu Rikugun Noborito Kenkyujo Shiryokan kansei ni atatte', www.meiji.ac.jp, 20 March 2010.
19. Oshima, 'Insatsu,' article 1, 32; Oshima, 'Insatsu,' article 2, 30.
20. Kitaoka Yasuo, 'Handai no sangakurenkei ga shakai o kaetta,' www.meti.go.jp, 17 February 2021; Okabe Kinjiro, https://ja.wikipedia.org.
21. Shimada, 'Kindai,' 75-76.
22. Kuba, 13; Nakamura, *Ichi*, 50-53; Nomura, 'Zaidan,' 69-71, 78. Nisseiken website, www.jp-nisseiken.co.jp.
23. Nakamura, *Ichi*, 215-216.
24. Kim, 'I,' 33-34; Yi, *Pukhan*, 220; Kim, 'Toi', 206.
25. Kim Chong-hui, 'Chisikindul un Tang kwa Suryong ui ryongdo mit e so man Choguk kwa inmin ul wihan chamdoen salm ul nurilsu

itta,' *Rodong Sinmun*, 11 December 1999; 'Destiny Changes,' *Korea Today*, August 2005, 14.

26. Nakamura, *Ichi*, 91; Tsuchiya, 'Dobutsu,' 13; Njeumi, 'Global,' 25.
27. Ri, 'Han,' 25.
28. Shiina, 'Gizo, 98, note 4.
29. Watanabe, '"Watakushi,"' 82. In marrying, Seki took her husband's name, Kobayashi.
30. Kinoshita, *Kesareta*, 292.
31. Saito, *Boryaku*, 13, 16-17.
32. Nagano/Akaho, *Kokosei*, 84. Noborito had developed the special poison, but the police interrogated many military veterans. One reason, according to a Noborito veteran, was that the poison had been widely distributed to military units near the end of the war for suicide. *See* 'Jintai jikken,' *Tokyo Shimbun*, 1998.08.14. Kitazawa Ryuji, responsible for the acetone cyanohydrin at war's end, had passed ampoules of the poison to AGS officers and other military men for suicide. He did not learn what happened to those ampoules. *See* Saito, *Boryaku*, 144; Nagano/Akaho, *Kokosei*, 115-16.
33. Saito, *Boryakusen*, 136. According to Takiwaki, acetone cyanohydrin would produce convulsions in victims in a few minutes and kill them in 15 to 30 minutes. He did not think the man sentenced for the murders, the artist Hirasawa Sadamichi, was guilty, viewing his knowledge of poison as 'zero.' Takiwaki did not think someone from Noborito had committed the crime but 'imagined' that it was a soldier. He also recalled that vials of acetone cyanohydrin had been among the items that the US Army Counter Intelligence Corps had confiscated from Noborito. *See* Saito, *Boryakusen*, 137. The Teikoku Bank (Teigin) incident attracted much attention, including that of the British novelist David Peace, who wrote of it in *Occupied City* (2011). He linked the incident to Unit 731 rather than to Noborito.
34. Kimura Manabu, 'Nisesatsu: Kokuhaku shi yatto otozureta shusen,' *Yomiuri Shimbun*, 23 August 1995; Kimura Manabu, 'Saikin to doku gasu: Yami no bunya, jujika omoku,' *Yomiuri Shimbun*, 30 August 1995.
35. 'Shinoda Ryo,' www.weblio.jp; Oshima, 'Insatsu,' article 2, 30; Hata, *Nihon*, 10; Ban, *Rikugun*, 204; 'Destiny Changes,' *Korea Today*, August 2005, 14.
36. Yamada, '8 gatsu,' 123-124.

Chapter 10: Remembering Noborito

1. Yamada and Meiji, *Rikugun*, 269.
2. Both the Army's Kaikosha and the Navy's Suikosha disbanded during the Occupation. Veterans revived both organisations after Japan recovered its sovereignty. Kaikosha is now open to members of Japan's postwar GSDF. The naval association, rebranded Suikokai, is open to members of the MSDF. Each organisation publishes its own periodical: *Kaiko* and *Suiko*. The Nakano Koyukai, the school's alumni group, published in 1978 a history of the Nakano School: *Rikugun Nakano Gakko*.
3. Tsukamoto, 'Tenji naiyo,' 3-4; Oshima, 'Insatsu,' second of two articles, 30.
4. Tsukamoto, 'Moto joinra,' 94. Watanabe, 'Noborito', 29.
5. Watanabe, 'Noborito,' 30. Watanabe, 'Watakushi', 87.
6. Tsukamoto, 'Tenji naiyo', 8.
7. Ban, 'Himitsu,' 97; Yamada, Watanabe, Tsukamoto, 'Noborito,' 112; Tsukamoto, '"Kai,"' 21.
8. Watanabe, '2015 nendo,' 2; Tsukamoto, 'Tenji naiyo,' 15.
9. Yamada, Meiji.
10. Meiji University, 'Shiryokan no tatemono wa itsu taterareta?' *Heiwa Kyoiku Noborito Kenkyujo Shiryokan Dayori*, 15 October 2010. The building housed projects for weapons to attack livestock and crops. Technician Matsukawa Hitoshi of Noborito's Second Section told a postwar researcher that he had worked in the building to develop a rice fungus that the Japanese Army dropped from aircraft onto Chinese rice fields. *See* Watanabe, 'Noborito,' 32.
11. Meiji University, '2010 nen 3 gatsu 29 nichi: Meiji Daigaku Heiwa Kyoiku Noborito Kenkyujo Shiryokan kaikan,' '4 gatsu 7 nichi yori ippan kokai,' *Heiwa Kyoiku Noborito Kenkyujo Shiryokan Dayori*, 21 July 2010. More than 92,000 visits by early May 2024, according to the museum. Communication of 7 May 2024 with Shiina Maho.
12. *See* Meiji, *Gaidobukku*, for a description of each exhibition room.
13. Meiji Daigaku Heiwa Kyoiku Noborito Kenkyujo Shiryokan, *10 nen no ayumi*, 6, 8, 9; Meiji Daigaku Heiwa Kyoiku Noborito Kenkyujo Shiryokan, 'Dai 13 kai.' Meiji, *Gaidobukku*, 4.

14. '*Meiji Daigaku Heiwa Kyoiku Noborito Kenkyujo Shiryokan' setsuritsu e,*' www.meiji.ac.jp/
15. Yamada Akira, 'Putting War Memories to Use for the Sake of the Future,' 24 February 2016, https://english-meiji.net/articles/37/.
16. Watanabe, 'Noborito,' 35; Yamada, Watanabe, and Tsukamoto, 'Noborito,' 110-111.
17. Kanagawa-ken, *Kanagawa-ken*, 68-69.
18. Kinoshita, *Kesareta*, 80; Meiji, *Gaidobukku*, 5; Saito, *Boryakusen*, 26.
19. Nakamura, *Ichi*, 102; "Our History," ROK Animal and Plant Quarantine Agency, www.qia.go.kr/english.
20. The UP news item appeared under various headlines, such as 'Balloons Didn't Work,' *Windsor Daily Star*, 16 August 1945; 'Japs Perfected Death Beam,' United Press, 7 October 1945; Ban and Shinoda, 'Noborito,' 687.
21. For information on the research areas of the 10 numbered ATRIs, *see* Grunden, *Secret*, 207.
22. Nagase-Reimer and Kawamura, 'Nihon,' 12.
23. Kusaba, 'Fusen,' 72-73; Sato, *Dai*, 283.
24. Kimura Manabu, 'Kuwairiki kosen: denpa heiki mo kaihatsu taisho,' *Yomiuri Shimbun*, 27 August 1995.Colonel Hayakawa Kiyoshi claimed that a Nisei US Army officer told him that he and others involved in biological warfare in the Second World War, presumably collaborating with the Americans in Occupied Japan, would be taken immediately to the United States in the event of war with the Soviet Union. *See* Yamada, 'Teigin,' 55.
25. Kusaba, 'Fusen,' 76. Fujiwara, *Ryukonroku*, 292.
26. Takahashi Kosuke, 'Japan Steps Up Development of High-Powered Microwave Weapons,' 25 January 2022, www.janes.com.

BIBLIOGRAPHY

For Japanese books, the place of publication is Tokyo unless otherwise noted.

Ariga Tsutao. *Nihon Riku-Kaigun no joho kiko to sono katsudo* [Intelligence Organs and Activities of Japan's Army and Navy]. Kindai Bungeisha, 1994.

Arisue Seizo. *Arisue Kikancho no shuki: shusen hishi* [Memoir of the Chief of the Arisue Agency: Secret History of thc War's End]. Fuyo Shobo Shuppan, 1987.

___. *Arisue Seizo kaikoroku* [Memoir of Arisue Seizo]. Fuyo Shobo, 1974.

___. *Seiji to gunji to jinji: Sanbo Honbu Daini Bucho no shuki* [Politics, Military Affairs, and Personnel: Memoir of the AGS Second Bureau Chief]. Fuyo Shobo, 1982.

Asahi Shinbunsha, ed. *Gendai jinbutsu jiten* [Contemporary Biographic Dictionary]. Asahi Shinbunsha, 1977.

Ban Shigeo. 'Himitsu heiki o tsukutta Noborito Kenkyujo' [The Noborito Research Institute That Made the Secret Weapons]. *Rekishi to jinbutsu*, October 1980, 96-103.

___. *Rikugun Noborito Kenkyujo no shinjitsu* [The Truth About the Army Noborito Research Institute]. Fuyo Shobo Shuppan, 2001.

Ban Shigeo and Shinoda Ryo. 'Noborito Kenkyujo no himitsu' [Secrets of Noborito Research Institute]. In *Rikusen Heiki Soran* [Overview of Land Warfare Weapons]. Edited by Nihon Heiki Kogyokai. Tosho Shuppansha, 1977.

Brophy, Leo P., Wyndham D. Miles, and Rexmond C. Cochrane. *The Chemical Warfare Service: From Laboratory to Field*. Washington, DC: Office of the Chief of Military History, Department of the Army, 1959.

Buxton, A. Carly. *Unthinking Collaboration: American Nisei in Transwar Japan*. Honolulu: University of Hawai'i Press, 2022.

Chong Il-gwon. *Chong Il-gwon hoegorok* [Memoir of Chong Il-gwon]. Seoul: Koryo Sojok: Kwangmyong Chulpansa, 1996.

Clark, Eugene Franklin. *The Secrets of Inchon: The Untold Story of the Most Daring Covert Mission of the Korean War*. New York: Putnam's, 2002.

Coen, Ross Allen. *Fu-go: The Curious History of Japan's Balloon Bomb Attack on America*. Lincoln, NB: University of Nebraska Press, 2014.

Cook, Haruko Taya, and Theodore F. Cook. *Japan at War: An Oral History*. New York: New Press, 1992.

Cunliffe, William H. *Select Documents on Japanese War Crimes and Japanese Biological Warfare, 1934-2006*. Washington, DC: National Archives and Records Administration, 2006. www.archives.gov/files/iwg/japanese-war-crimes/select-documents.pdf.

Drea, Edward J., et al. *Researching Japanese War Crimes Records: Introductory Essays*. Washington, DC: Nazi War Crimes and Japanese Imperial Government Records Interagency Working Group, 2006.

Faligot, Roger, and Pascal Krop. *La piscine: les services secrets français, 1944-1984*. Paris: Editions du Seuil, 1985.

Fang Jianchang, 'Senchu, Jukei seifu ni taisuru Nihon no nisesatsu kosaku' [Japan's Wartime Counterfeiting Operation Against the Chungking Government]. *Chugoku tsushin*, Spring 2002, 86-91.

Friedman, Herbert A. 'Propaganda Banknotes of the Vietnam War,' www.psywarrior.com/Vietnambanknote.html.

Fujiwara Iwaichi. 'Boryakuka' [Clandestine Operations Section]. *Rekishi to jinbutsu*, August 1985, 68-72.

___. *Ryukonroku* [Record of the Spirit That Remains]. Urawa: Shingaku Shuppan, 1986.

Gehlen, Reinhard. *The Service: The Memoirs of General Reinhard Gehlen*. Translated by David Irving. New York: World Publishing Company, 1972.

Gold, Hal. *Unit 731: Testimony*. Tokyo: Yenbooks, 1996.

Goto Masao. 'Genbaku to kessen himitsu heiki' [The Atomic Bomb and Decisive Secret Weapons]. *Shukan Yomiuri*, special issue, 8 December 1956, 178-186.

Grant, Ben J. and Robert V. Guelich. 'Air War in the Pacific.' *Air Force*, December 1945, 3-34.

Grunden, Walter E. *Secret Weapons and World War II: Japan in the Shadow of Big Science*. Lawrence, KS: University Press of Kansas, 2005.

Hall, Roger. *You're Stepping on My Cloak and Dagger*. Annapolis, MD: Naval Institute Press, 2004.

'Harubin Tokumu Kikan kaimetsusu: nokosareta hikiage e no moten' [Destroy the Harbin Special Service Agency. Blind Spot for Those Left Behind]. *Shukan Yomiuri*, 7 March 1954, 44-48.

Harris, Sheldon H. *Factories of Death: Japanese Biological Warfare, 1932-1945, and the American Cover-up*. Rev. ed. New York: Routledge, 2002.

Hata Ikuhiko, ed. *Nihon Riku-Kaigun sogo jiten* [Comprehensive Dictionary of the Japanese Army and Navy]. 2nd ed. Tokyo Daigaku Shuppankai, 2005.

Hatakeyama Seiko. *Hiroku Rikugun Nakano Gakko* [Secret Record of the Army Nakano School]. Edited by Hosaka Masayasu. Shinchosha, 2003.

Hatakeyama Seiko. *Rikugun Nakano Gakko shusen hishi* [Army Nakano School, Secret History of the War's End]. Edited by Hosaka Masayasu. Shinchosha, 2004.

Hori Eizo. *Daihonei sanbo no joho senki: Joho naki kokka no higeki* [Intelligence Warfare Record of an IGHQ Staff Officer: The Tragedy of a Country Without Intelligence]. Bungei Shunju, 1989.

Houghton, Vince. *Nuking the Moon: And Other Intelligence Schemes and Military Plots Left on the Drawing Board*. New York: Penguin Books, 2019.

Howard, Michael. *British Intelligence in the Second World War*, Volume 5, *Strategic Deception*. New York: Cambridge University Press, 1990.

Hunt, E. Howard, with Greg Aunapu. *American Spy: My Secret History in the CIA, Watergate, and Beyond*. Hoboken, NJ: John Wiley & Sons, 2007.

Im Chong-hyok, editor. *Gendai Chosen no kagakushatachi* [Scientists of Modern Korea]. Sairyusha, 1997.

Imai Takeo. *Showa no boryaku* [Showa Clandestine Operations]. Asahi Sonorama, 1985.

Inaba Masao. *Boryaku shizai seizo kojo* [Clandestine Operations Material Production Plant]. *Shukan Yomiuri*, special issue, 8 December 1956, 27-30.

Iwakuro Hideo. *Iwakuro Hideo shi danwa sokkiroku* [Fast Record of a Conversation with Iwakuro Hideo]. Interview conducted by Kido Nikki Kenkyukai. Nihon Kindai Shiryo Kenkyukai, 1977.

___. 'Iwakuro Kikan shimatsuki' [Record of the Course of Events for the Iwakuro Agency]. *Shukan Yomiuri*, special issue, 8 December 1956, 118-121.

___. 'Junbi sareta himitsusen' [Clandestine Warfare Prepared]. *Shukan Yomiuri*, special issue, 8 December 1956, 16-26.

___. *Kagaku jidai kara ningen no jidai e* [From an Era of Science to an Era of Man]. Risosha, 1970.

___. *Sensoshiron* [On the History of Warfare]. Koseisha Koseikaku, 1967.

___. *Shingaporu sokogeki: Konoe Hohei Daigo Rentai dengeki senki* [General Offensive on Singapore: Record of the Blitzkrieg of the 5th Regiment, Imperial Guards]. Kojinsha, 2000.

Kahn, Herman, and Iwakuro Hideo. 'Shukyo to kagaku no mirai (taidan)' [The Future of Religion and Science (Dialogue)]. *Jiyu*, February 1969, 106-119.

Kanagawa-ken Rekishi Kyōikusha Kyōgikai, ed. *Kanagawa-ken no senso iseki* [The Remains of War in Kanagawa Prefecture]. Otsuki Shoten, 1996.

Kato Masao. *Rikugun Nakano Gakko: himitsusenshi no jittai* [The Army Nakano School: The Truth About the Covert Warriors]. Kojinsha, 2001.

Kawabe Torashiro. *Kawabe Torashiro kaisoroku: Ichigayadai kara Ichigayadai e* [Memoir of Kawabe Torashiro: From Ichigayadai to Ichigayadai]. Mainichi Shinbunsha, 1979.

Kawai Masahiro. 'Ichi gunjin no sengo: Iwakuro Hideo to Kyoto Sangyo Daigaku' [A Military Man After the War: Iwakuro Hideo and Kyoto Sangyo University]. *Sandai Hogaku*, April 2017, 27-43.

Kawashima Hideo. *Ushi no byoki to sono yoboho* [Cattle diseases and their prevention]. Seikatsusha, 1944.

Kikuchi Katsuhiro. 'Kindai kenchikushi kara Noborito Kenkyujo o yomitoku' [Deciphering the Noborito Research Institute from the Vantage Point of Architecture's Modern History]. *Meiji Daigaku Heiwa Kyoiku Noborito Kenkyujo Shiryokan Kanpo*, 30 September 2020, 137-170.

Kim Chong-hui (Kim Jong Hui). 'I Found the Real Way to Scientific Pursuits under the Wing of the Fatherly Leader.' *Korea Today*, December 1976, 33-35.

___. 'Toi michi o aruitekita jinsei' [My Life, A Long Path Walked]. In Im Chong-hyok, editor, *Gendai Chosen no kagakushatachi* [Scientists of Modern Korea]. Sairyusha, 1997.

Kinoshita Kenzo. *Kesareta himitsusen kenkyujo* [Erased Clandestine Warfare Research Institute]. Nagano: Shinano Mainichi Shinbunsha, 1994.

___. 'Nagano-ken ni okeru Rikugun Noborito Kenkyujo no sokai shiryo ni tsuite: Kami-Ina Chiho (Inadani) o chushin to shite' [On Materials

Related to the Evacuation of the Army Noborito Research Institute to Nagano Prefecture: Centring on the Kami-Ina Area (Ina Valley)]. *Meiji Daigaku Heiwa Kyoiku Noborito Kenkyujo Shiryokan Kanpo*, 30 September 2016, 3-21.

Kojunsha, ed. *Nihon shinshiroku* [Who's Who in Japan, all editions]. Tokyo: Kojunsha Shuppankyoku.

Kotani Ken. *Japanese Intelligence in World War II.* Translated by Kotani Chiharu. Oxford, UK; New York: Osprey, 2009.

___. *Nihongun no interijensu: naze joho ga ikasarenai no ka* [Japanese Military Intelligence: Why Was Intelligence Not Used?]. Kodansha, 2007.

Kuba Noboru. *Rikugun Dai Kyu Gijutsu Kenkyujo Dai Shichi Kenkyuhan kenkyu gaiyo: fusen bakudan tosai gyueki funmatsu byodoku o mote suru taibei kogeki* [An Outline of Research Group 7, Army 9th Technical Research Institute: Attacking the United States by Means of Powdered Rinderpest Virus Loaded on Bombing Balloons]. 31 March 1990. Unpublished. Copy on file with thc Dcfunct Imperial Japanese Army Noborito Laboratory Museum for Education in Peace.

Kumano Sanpei. *'Sakata Kikan' Shutsudosu: shirarezaru tai-Shi choho kosaku no uchimaku* [The 'Sakata Agency' Sets Forth: Behind the Scenes of Unknown Clandestine Operations Against China]. Tendensha, 1989.

Kusaba Sueki. 'Fusen bakudan ni yoru Beihondo kogeki' [Attacking the US Mainland with Bombing Balloons]. In *Rikusen heiki soran* [Overview of Land Warfare Weapons]. Edited by Nihon Heiki Kogyokai. Tosho Shuppansha, 1977.

___. 'Fusen bakuden: sono aidea to iryoku no subete' [Bombing Balloons: Everything About the Idea and Its Force]. *Maru*, November 1961, 72-76.

___. 'Tokushu heiki kenkyu no zenbo' [Everything About Special Weapons Research]. In *Rikusen heiki soran* [Overview of Land Warfare Weapons]. Edited by Nihon Heiki Kogyokai. Tosho Shuppansha, 1977.

Kusuyama Tadayuki. *'Rikugun Noborito Kenkyujo' o toru* [Filming *The Army Noborito Research Institute*]. Fujinsha, 2014.

Kuwahara Takeshi. "Nakano Gakko' [The Nakano School]. In *Showa Gunji Hiwa: Dodai Kurabu koenshu* [Showa Military Secret Stories – Dodai Club Collected Talks]. Volume 1. Edited by 'Dodai Club Collected Talks' Editorial Board. Dodai Economic Club, 1987.

Kye In-ju. *Maegado Changgun kwa Kye In-ju Taeryong* [General MacArthur and Colonel Kye In-ju]. Seoul: Tain Midio, 1999.

Kyu shokuminchi jinji soran: Chosen hen [Personnel Overview of Former Colonies: Chosen]. Volume 8. Nihon Tosho Centre, 1997.

LaPaz, Lincoln, with Albert Rosenfeld, 'Now It Can Be Told: Japan's Balloon Invasion of America.' *Collier's*, 17 January 1953, 9-11.

Leverkuehn, Paul. *German Military Intelligence*. Translated by R.H. Stevens and Constantine FitzGibbon. London: Weidenfeld and Nicholson, 1954.

Lisle, John. *The Dirty Tricks Department: Stanley Lovell, the OSS, and the Masterminds of World War II Secret Warfare*. New York: St. Martin's Press, 2023.

Lovell, Stanley P. *Of Spies & Stratagems*. Englewood Cliffs, NJ: Prentice Hall, 1963.

Maema Takanori. *Fugaku: Bei hondo o bakugekiseyo*, [Fugaku: Bomb the Continental United States]. Volumes 1 and 2. Kodansha, 1995.

Mainichi Shinbun Tokubetsu Hodobu Shuzaihan. *Okinawa Senso mararia jiken: minami no shima no kyosei sokai* [Malaria Incident in the Battle for Okinawa: Forced Evacuation of a Southern Island]. Osaka: Toho Shuppan, 1994.

Martin, Brian G. 'Shield of Collaboration: The Wang Jingwei Regime's Security Service, 1939-1945.' *Intelligence and National Security*, Winter 2001, 89-148.

Mashbir, Sydney. *I Was an American Spy*. New York: Vantage Press, 1953.

Matsudaira Michio. *Kindai kagakusen* [Modern Scientific Warfare]. Nihon Koronsha, 1940.

Matsuno Seiya. 'Nihon Rikugun no himitsusen kizai: bocho kizai, choho kizai, boryaku kizai, senden kizai no jittai' [The Japanese Army's Clandestine Warfare Equipment – The True State of the Counterintelligence, Intelligence, Clandestine Operations, and Propaganda Equipment]. *Meiji Daigaku Heiwa Kyoiku Noborito Kenkyujo Shiryokan Kanpo*, 30 September 2020, 1-36.

McVety, Amanda Kay. *The Rinderpest Campaigns: A Virus, Its Vaccines, and Global Development in the Twentieth Century*. Cambridge, UK; New York: Cambridge University Press, 2018.

Meiji Daigaku Heiwa Kyoiku Noborito Kenkyujo Shiryokan. *Dai 13 kai kikakuten 'Himitsu kikan "Yama Kikan" to Noborito Kenkyujo – Nihon Rikugun no bocho to wa, Zoruge Jiken 80 nen' kaisai no annai* [Announcement of Opening: 13th Exhibition, 'Clandestine Organisation "Yama Agency" and the Noborito Research Institute – What Was Japanese Army Counterintelligence? 80 Years After the Sorge Incident'] www.meiji.ac.jp/noborito/info/2022/mkmht0000001sl9j.html

Meiji Daigaku Heiwa Kyoiku Noborito Kenkyujo Shiryokan. *10 nen no ayumi* [A Course of 10 Years]. 29 March 2020.

Meiji Daigaku Heiwa Kyoiku Noborito Kenkyujo Shiryokan gaidobukku (A Guide to the Defunct Imperial Japanese Army Noborito Laboratory Museum for Education in Peace). 5th ed., Kawasaki: 2019.

Melton, H. Keith. *Ultimate Spy*. New York: DK Publishing, 2015.

Mercado, Stephen C. Review of *Rikugun Noborito Kenkyujo no shinjitsu* [The Truth About the Army Noborito Research Institute], by Ban Shigeo. *Studies in Intelligence*, December 2002, 79-81.

___. 'The Japanese Army's Noborito Research Institute.' *International Journal of Intelligence and CounterIntelligence*, Summer 2004, 286-299.

___. *The Shadow Warriors of Nakano: A History of the Imperial Japanese Army's Elite Intelligence School*. Washington, DC: Brassey's, 2002.

Mikesh, Robert C. *Japan's World War II Balloon Bomb Attacks on North America*. Washington, DC: Smithsonian Institution Press, 1973.

Miyamoto Masaki. 'Enka biniru gijutsushi no gaiyo to shiryo chosa kekka (2) [An Outline of the Technical History of Polyvinyl Chloride and Results of Documentary Research (2)]. In National Museum of Science and Technology, ed. *Kokuritsu Kagaku Hakubutsukan gijutsu no keitoka chosa kekka* [National Museum of Science and Technology Research on the Systematisation of Technology]. Volume 2. National Museum of Science and Technology, 2002.

Mizutani Naoko. '1644 Butai no soshiki to katsudo' [Organisation and Activities of Unit 1644]. *Senso sekinin kenkyu*, Spring 1997, 50-59.

Nagano/Akaho High School Peace Seminar, Kanagawa/Hosei University Daini Junior & Senior High School Peace Research Society. *Kokosei ga ou Rikugun Noborito Kenkyujo* [High School Students in Search of Army Noborito Research Institute]. Kyoiku Shiryo Shuppankai, 1991.

Nagase-Reimer, Keiko, and Kawamura Yutaka. 'Nihon ni okeru kyoryoku denpa heiki kaihatsu keikaku no keifu: senjika no "satsujin kosen" ni kansuru kento' [Lineage of Development Plans in Japan for Powerful Electric-Wave Weapons: Examining the 'Death Ray' in the War]. *Il Saggiatore*, 2014, 1-16.

Nakamura Junji. *Ichi jueki kenkyusha no ayumi* [An Epizootic Researcher's Course]. Iwanami Shoten, 1975.

Nakano Koyukai, ed. *Rikugun Nakano Gakko* [Army Nakano School]. Nakano Koyukai, 1978.

Njeumi, Felix, 'Global rinderpest eradication: The Achievements of FAO and GREP.' In *Lessons Learned from the Eradication of Rinderpest for Controlling Other Transboundary Animal Diseases*, GREP Symposium and High-Level Meeting, 12-15 October 2010. www.fao.org.

Nomura Yoshitoshi. 'Zaidan Hojin Nihon Seibutsu Kagaku Kenkyujo no rekishi' (History of Nippon Institute for Biological Science). *Nihon jui shigaku zasshi*, February 2017, 66-79.

Nouzille, Vincent, with Olivier Huwart. 'Comment la France a recruté des savants de Hitler' [How France Recruited Some of Hitler's Scientists]. *Express international*, 20-26 May 1999, 52-64.

Okada Yoshimasa. 'Chugoku shihei gizo jiken no zenbo' [Everything About the Chinese Paper Note Counterfeiting Incident]. *Rekishi to jinbutsu*, October 1980, 42-51.

___. 'Shina Jihen ni okeru keizai boryaku' [Economic Clandestine Operations in the China Incident]. In *Showa Gunji Hiwa: Dodai Kurabu Koenshu*, Volume 1 [Showa Military Secrets: Volume 1, Collected Speeches of the Dodai Economic Club]. Edited by Dodai Economic Club Speech Collection Editorial Committee. Dodai Economic Club, 1987.

Okubo Takeo. *Uminari no hibi: kakusareta sengoshi no danso* [Days of Rumbling Seas: Hidden Gap in Postwar History]. Kaiyo Mondai Kenkyukai, 1978.

Onoda Hiroo. *No Surrender: My Thirty-Year War*. Translated by Charles S. Terry. Tokyo; New York: Kodansha International, 1974.

Oshima Yasuhiro. 'Insatsu gijutsu no saikoho, nisesatsu o tsukure: Rikugun Gijutsu Kenkyujo no omoide (Zenpen)' [The Highest Peak of Printing Technology, Making Counterfeit Notes: Memories of an Army Technical Research Institute (First Part)]. Article 1 of 2. *Insatsu zasshi*, November 2006, 31-35.

___. 'Insatsu gijutsu no saikoho, nisesatsu o tsukure: Rikugun Gijutsu Kenkyujo no omoide (Kohen)' [The Highest Peak of Printing Technology, Making Counterfeit Notes: Memories of an Army Technical Research Institute (Concluding Part)]. Article 2 of 2. *Insatsu zasshi*, December 2006, 27-30.

Ota Masakatsu. *731 menseki no keifu: Saikinsen butai to hizo no fairu* [Lineage of 731 Immunity: A Bacteriological Warfare Unit and the Prized Files]. Nihon Hyoronsha, 1999.

Otake Osamu. 'Nihon no kindai juigaku shi: Gakujutsukaigicho o tsutometa juikai no kyosei, Ochi Yuichi' (Modern History of Veterinary Medicine in Japan – Dr Yuichi Ochi: A Giant Star of Veterinary

Science World, Served as the President of the Science Council of Japan). *Dobutsu rinsho igaku,* No. 3, 2016, 114-118.

Paek Son-yop (Paik Sun Yup). *From Pusan to Panmunjom*. Washington: Brassey's, 1992.

Petersen, Michael. 'The Intelligence That Wasn't: CIA Name Files, the US Army, and Intelligence Gathering in Occupied Japan.'In Edward J. Drea et al., *Researching Japanese War Crimes: Introductory Essays*. Washington, DC: Nazi War Crimes and Japanese Imperial Government Records Interagency Working Group, 2006.

Pierard, Burt. 'Japanese Balloon Bombs and the Hanford Engineer Works (Re-visited),' *The Moderator*, January – March 2021, 3-4.

Ri Kyong-hui. 'Han kwahakcha ui iyagi' [The Story of a Scientist], *Kumsu Kangsan*, December 2004, 25.

Saito Michinori. *Boryakusen: dokyumento Rikugun Noborito Kenkyujo* [Clandestine Warfare: Documenting the Army Noborito Research Institute]. Jiji Tsushinsha, 1987.

___. *Boryakusen: Rikugun Noborito Kenkyujo* [Clandestine Warfare: The Army Noborito Research Institute]. Gakushu Kenkyusha, 2001.

___. *Maboroshi no tokumu kikan 'Yama': Showashi hakkutsu* [Digging Up Showa History: 'Yama,' the Phantom Special Service Agency:]. Shinchosha, 2003.

___. *Shogen Rikugun Nakano Gakko: sotsugyoseitachi no tsuiso* [Testimonies, Army Nakano School: Reminiscences of Graduates]. Basilico, 2013.

___. *Supai akademi Rikugun Nakano Gakko: senjichu, nananenkan dake Nihon ni sonzaishita 'himitsu chohoin yosei kikan' no subete* [Spy Academy, the Army Nakano School: Everything About the 'Training Organ for Covert Operatives' That Existed for Only Seven Years in Wartime Japan]. Yosensha, 2015.

___. 'Zoruge o toraeta maboroshi no tokumu kikan "Yama."' ['Yama,' the Phantom Agency That Apprehended Sorge]. *Bungei Shunju*, August 2003, 338-346.

Samuels, Richard J. *Special Duty: A History of the Japanese Intelligence Community*. Ithaca, NY: Cornell University Press, 2019.

Satake Kinji. 'Denpa heiki no zenbo' [Everything About Electric Wave Weapons]. In *Rikusen heiki soran* [Overview of Land Warfare Weapons]. Edited by Nihon Heiki Kogyokai. Tosho Shuppansha, 1977.

Sato Kenryo. *Dai Toa Senso kaikoroku* [Memoir of the Greater East Asian War]. Tokuma Shoten, 1966.

Seydewitz, Max, and Kurt Karl Doberer. *Todesstrahlen und andere neue Kriegswaffen* [Death Rays and Other New Weapons of War]. London: Malik, 1936.

___. *Les Rayons de la mort et autres nouveaux engins de guerre*. [Death Rays and Other New Weapons of War]. Translated and adapted by Captain G.P. Capart. Paris: Hachette, 1937.

___. Kyoi no shin heiki satsujin kosen [Astonishing New Weapon: The Death Ray]. Translated by Kakimoto Ryohei. Sanpo Shuppansha, 1940.

Shiina Maho, 'Gizo shihei ni riyo sareta "kami"' [The 'Paper' Used in the Counterfeit Bank Notes], *Meiji Daigaku Heiwa Kyoiku Noborito Kenkyujo Shiryokan Kanpo*, 25 March 2016, 73-100

___. '"Noborito Jikkenba," "Himitsusen Shizai Kenkyushitsu" jidai no doin to "Noborito Shutchojo" kaisetsu made' [Mobilisation From the Time of the 'Noborito Laboratory' and 'Clandestine Warfare Materials Laboratory' to the Establishment of the Noborito Branch Depot]. *Meiji Daigaku Heiwa Kyoiku Noborito Kenkyujo Shiryokan Kanpo*, 1 September 2018, 31-53.

___. 'Tenji' [Exhibition]. *Meiji Daigaku Heiwa Kyoiku Noborito Kenkyujo Shiryokan Kanpo*. 30 September 2020, 37-67.

Shimada Moriya. 'Kindai kishogaku no kaitakusha Arakawa Hidetoshi ryakuden' [Brief Biography of Arakawa Hidetoshi, a Pioneer in Modern Meteorology], *Tenki*, December 1996, 75-76.

Shinta Shoji. *Hiwa Rikugun Noborito Kenkyujo no seishun* [A Secret Story: My Youth in the Army Noborito Research Institute]. Kodansha, 2004.

Snow, Philip. *The Fall of Hong Kong: Britain, China, and the Japanese Occupation*. New Haven: Yale University Press, 2003.

Sun Tzu. *The Art of War*. Translated by Samuel B. Griffith. New York: Oxford University Press, 1963.

Suzuki Shunpei. *Fusen bakudan: saigo no kessen heiki* [The Balloon Bomb: The Final Decisive Weapon]. Kojinsha, 2001.

Tada Reikichi, ed. *Kokubo gijutsu* [National Defence Technology]. Hakuyosha, 1942.

___. *Shoraisen to kagaku shinheiki* [Future Warfare and New Weapons of Science]. Shin Toa Kyokai, 1942.

Takai Saburo, 'Saitenken!: Jieitai no joho kikan' [Re-examination!: Intelligence Organs of the Self-Defence Forces], *Gunji Kenkyu*, July 1997, 192-209.

Tanaka Toshio. *Rikugun Nakano Gakko no Tobu Nyu Ginia yugekisen: Taiwan Takasagohei to no senshoroku* [Army Nakano School's

Guerrilla Warfare in Eastern New Guinea: A Record of Victories with Taiwan Natives]. Senshi Kankokai, 1996.

Torii Hideharu. *Nihon Rikugun no tsushin chohosen: Kitatama Tsushinjo no bojushatachi* [The Japanese Army's Communications Intelligence War: The Interceptors of Kitatama Communications Station]. Tachikawa: Keyaki Shuppan, 2011.

Toyama Misao and Joho Yoshio, eds. *Riku-Kaigun shokan jinji soran* [Personnel Guide to Officers of the Imperial Army and Navy: Army]. Fuyo Shobo, 1981.

Toye, Hugh. *The Springing Tiger. A Study of a Revolutionary*. London: Cassell, 1959.

Toynbee, Arnold, and Iwakuro Hideo. '21 seki no sekai (taidan)' [The World in the 21st Century (Dialogue)]. *Jiyu*, March 1968, 38-54.

Turner, Des. *Aston House: Station 12 – SOE's Secret Centre*. Stroud: Sutton Publishing, 2006.

___. *The Secrets of Station 14: Briggens House, SOE Forgery and Polish Elite Agent Training Station*. Cheltenham: Thc History Press, 2022

Tsuchiya Kotaro. 'Dobutsuyo wakuchin no hensen to kansestusho no seiatsu·haijo·konzetsu: Nihon no gyueki wakuchin to kyokenbyo wakuchin o chushin ni' (Progresses in Veterinary Vaccines for the Control, Elimination, and Eradication of Animal Infectious Diseases: Rinderpest Vaccines and Rabies Vaccines Developed in Japan). *Nihon juishigaku zasshi*, February 2012, 10-13, 10–21.

Tsukamoto Yuriko. 'Fusen bakudan sakusen no suiko to shuketsu' [Implementation and End of the Bombing Balloon Operation], *Meiji Daigaku Heiwa Kyoiku Noborito Kenkyujo Shiryokan Kanpo*, 30 September 2016, 23-41.

___. '"Kai sosa shuki" yori akiraka ni natta kyu Nihon Rikugun no dokubutsu kenkyu to nettowaku oyobi GHQ to kawasareta gibu-ando-teiku' [Poison Research and Network of the Former Imperial Japanese Army, Revealed by 'Chief Inspector Kai's Notes,' and the 'Give and Take' with GHQ], *Meiji Daigaku Heiwa Kyoiku Noborito Kenkyujo Shiryokan Kanpo*, 25 September 2019, 1-26.

___. 'Moto joinra no sengo,' [The Postwar Lives of Former Noborito Research Institute Members], *Meiji Daigaku Heiwa Kyoiku Noborito Kenkyujo Shiryokan Kanpo*, 30 September 2016, 83-100.

___. 'Rikugun Kagaku Kenkyujo kara miru kagaku gijutsu no senso doin' [Mobilisation in War of Science and Technology Seen from the Vantage Point of the Army Noborito Research Institute]. *Meiji Daigaku*

Heiwa Kyoiku Noborito Kenkyujo Shiryokan Kanpo, 1 September 2018, 15-29.

___. *Tenji* [Exhibition], *Meiji Daigaku Heiwa Kyoiku Noborito Kenkyujo Shiryokan Kanpo*, 30 September 2023, 1-33.

___. 'Tenji naiyo kaisetsu' [An Explanation of the Exhibition's Contents]. *Meiji Daigaku Heiwa Kyoiku Noborito Kenkyujo Shiryokan Kanpo*, 30 September 2021, 1-19.

___. 'Yokosuka ni atta gokuhi kikan: Rikugun Noborito Kenkyujo to Yokouska [The Top-Secret Organisation at Yokosuka: The Army Noborito Research Institute and Yokosuka], *Meiji Daigaku Heiwa Kyoiku Noborito Kenkyujo Shiryokan Kanpo*, 30 September 2020, 127-135

Tsuneishi Keiichi and Asano Tomizo. *Saikinsen butai to jiketsushita futari no igakusha* [Bacteriological Warfare Unit and the Suicide of Two Physicians]. Shinchosha, 1982.

UEC Museum, University of Electro-Communications. 'III. Vacuum Tube,' 'Excerpt from NEC Total Engineering of Electrical Communication,' February 1954. www.museum.uec.ac.jp

Unno Juza. *Tokyo Kubaku: Gunjin shosetsushu* [The Bombing of Tokyo: A Collection of Soldier Stories]. Rajio Kagakusha, 1938.

Watanabe Kenji. '2015 nendo Kawasaki-shi Bunkasho jusho no kotoba' [Remarks on Receiving the 2015 Kawasaki City Cultural Award]. *Meiji Daigaku Heiwa Kyoiku Noborito Kenkyujo Shiryokan Kanpo*, 30 September 2016, 1-2.

___. 'Noborito Kenkyujo horiokoshi undo 30 nen no ayumi' [Thirty Years of Unearthing the Noborito Research Institute]. *Meiji Daigaku Heiwa Kyoiku Noborito Kenkyujo Shiryokan Kanpo,* 30 September 2021, 21-39.

___. 'Rikugun Noborito Kenkyujo no jisso o mitsumete: Meiji Daigaku Heiwa Kyoiku Noborito Kenkyujo Shiryokan no igi [Looking at the Reality of the Army Noborito Research Institute: The Meaning of Meiji University's Defunct Imperial Japanese Army Noborito Laboratory Museum for Education in Peace]. *Meiji Daigaku Heiwa Kyoiku Noborito Kenkyujo Shiryokan Kanpo*, 30 September 2023, 70-74.

___. '"Watakushi no machi kara senso ga mieta" – Noborito Kenkyujo ni kinmu shita shojo to no deai' ['War Was Visible From My Town' – Encounter With a Young Woman Who Worked at the Noborito Research Institute], *Meiji Daigaku Heiwa Kyoiku Noborito Kenkyujo Shiryokan Kanpo*, 30 September 2020, 79-104.

Watanabe Kenji and Yamada Akira. 'Moto fusen bakudan seizo doin joshi seito ni yoru shogenkai,' [Session of Testimony by Female Former

Students Mobilised To Make Bombing Balloons]. *Meiji Daigaku Heiwa Kyoiku Noborito Kenkyujo Shiryokan Kanpo*, 1 September 2018, 85-106.

Wells, H.G. *The War of the Worlds*. New York and London: Harper & Brothers, 1898.

Wexler, Philip, ed. *Toxicology in the Middle Ages and Renaissance*. London: Academic Press, 2017.

Wilbur, W.H. 'Those Japanese Balloons.' *Reader's Digest*, August 1950, 23-26.

Willoughby, Charles Andrew. *GHQ shirarezaru chohosen: Wirobi kaikoroku* [GHQ's Unknown Intelligence War: The Willoughby Memoir]. Edited by Yon C hong (En Tei) and Hiratsuka Masao. Yamakawa Shuppansha, 2011.

___. *Maneuver in War*. Harrisburg, PA: The Military Service Publishing Co., 1939.

Yamada Akira, 'Dai ni ki: 8 gatsu 15 nichi iko no Noborito Kenkyujo.' [The Noborito Research Institute After 15 August]. *Meiji Daigaku Heiwa Kyoiku Noborito Kenkyujo Shiryokan Kanpo*, 30 September 2016, 121-139.

___. 'Kagaku gijutsu to minkanjin no senso doin: Rikugun Noborito Jikkenba kaisetsu 80 nen' [The Mobilisation in War of Science, Technology, and Civilians: Eighty Years Since the Opening of the Army Noborito Laboratory]. *Meiji Daigaku Heiwa Kyoiku Noborito Kenkyujo Shiryokan Kanpo*, 1 September 2018, 59-83.

___. 'Kami to senso: Noborito Kenkyujo to fusen bakudan, nisesatsu' [Paper and War: Noborito Research Institute and Bombing Balloons, Counterfeit Bank Notes]. *Meiji Daigaku Heiwa Kyoiku Noborito Kenkyujo Shiryokan Kanpo*, 25 March 2016, 101-123.

___. 'Meiji Daigaku ni nokoru senso iseki (2)' [What Remains from the War at Meiji University (2)], Meiji Daigakushi Shiryo Senta, February 2022, www.meiji.ac.jp/

___. 'Teigin Jiken to Rikugun Noborito Kenkyujo: sosa shuki kara akiraka ni naru kyu Nihon Rikugun no dokubutsu kenkyu [The Teikoku Bank Incident and the Army Noborito Research Institute: The Former Japanese Army's Poison Research Revealed in Investigation Notes]. *Meiji Daigaku Heiwa Kyoiku Noborito Kenkyujo Shiryokan Kanpo*, 25 September 2019, 27-57.

Yamada Akira and Meiji Daigaku Heiwa Kyoiku Noborito Kenkyujo Shiryokan, eds. *Rikugun Noborito Kenkyujo 'himitsusen' no sekai:*

fusen bakudan, seibutsu heiki, nisesatu o saguru [Army Noborito Research Institute's World of 'Secret Warfare': The Search for Balloon Bombs, Biological Weapons, and Counterfeit Bank Notes]. Meiji Daigaku Shuppankai, 2012.

Yamada Akira and Watanabe Kenji, 'Eiga "Teikoku Jiken Shikeishu" tokusho' [Discussion on the Film *Teikoku Incident, Death Row Inmate*]. Moderated by Tsukamoto Yuriko. *Meiji Daigaku Heiwa Kyoiku Noborito Kenkyujo Shiryokan Kanpo*, 25 September 2019, 201-210.

Yamada Akira, Watanabe Kenji, and Tsukamoto Yuriko. 'Noborito Kenkyujoin ga katatta Teikoku Jiken to sono kensho' [What Noborito Research Institute Personnel Said about the Teikoku Bank Incident and Its Verification]. *Meiji Daigaku Heiwa Kyoiku Noborito Kenkyujo Shiryokan Kanpo*, 30 September 2020, 105-126.

Yamada Sakura. *Kagaku heiki* [Chemical Weapons]. Kyoritsusha, 1935.

Yamamoto Kenzo. '*Nise hohei kosaku no tenmatsu*,' [Details of the Counterfeit Fapi Operation]. *Rekishi to jinbutsu*, August 1985, 103-106.

___. *Rikugun nisesatsu sakusen: keikaku· jikkosha ga akasu Nitchusen hiwa* [Army Counterfeit Bank Note Operation: Secret Story of the War Between Japan and China Revealed by the Man Who Planned and Executed the Operation]. Gendaishi Shuppankai, 1984.

Yamamoto Taketoshi. *GHQ no kenetsu, choho, senden kosaku* [GHQ's Censorship, Intelligence, and Propaganda Operations]. Iwanami Shoten, 2013.

___. *Kenetsukan: hakken sareta GHQ meibo* [Censors: The Discovered GHQ Name List]. Shinchosha, 2021.

Yamamoto Yoshio. 'Seishun o kaketa gyokai' [The Industry To Which I Staked My Youth]. *Kanto Shibuho*, Japan Association of Surveyors, No. 41, 2012, 7-8.

Yanagida Yukiko. *Nisei heishi gekisen no kiroku: Nikkei Amerikajin no Dainiji Taisen* [Nisei Soldiers, A Record of Hard Fighting: Second-Generation Japanese Americans in the Second World War]. Shinchosha, 2012.

Yardley, Herbert O. *The American Black Chamber*. Indianapolis: Bobbs-Merrill, 1931.

___. *Burakku Chenba: Beikoku wa ika ni shite gaiko hiden o nusunda ka?* [The American Black Chamber: How Did the United States Steal Secret Diplomatic Telegrams?]. Translated by Osaka Mainichi Shinbunsha. Osaka: Osaka Mainichi Shinbunsha, 1931.

___. *Secret Service in America: The American Black Chamber*. London: Faber and Faber, 1940.

Yi Chae-sung. *Pukhan ul umjiginun tekunokuratu: Pukhan kwollyok ui silse; kwahakchadul ul almyon Pukhan i poinda* [Technocrats Who Are Moving North Korea: North Korean Real Power: North Korea Is Visible If You Know the Scientists]. Seoul, Ilpit, 1998.

Yi Chang-gon. *KLO ui Hangukchon pisa* [A Secret History of the KLO in the Korean War]. Seoul: Chisongsa, 2005.

Yick, Joseph K.S. 'Communist-Puppet Collaboration in Japanese-Occupied China: Pan Hannian and Li Shiqun, 1939-43,' *Intelligence and National Security*, Winter 2001, 61-88.

Yon Chong (En Tei). *Kyanon Kikan kara no shogen* [Testimony from the Canon Agency]. Bancho Shobo, 1973.

Yoshida Yoko. 'Pasokon tsushin de wakamono ni kataritsugu "senso hiwa", [Passing Down to Young People 'Stories of Secret War' Via Computer Communication], *Gendai*, November 2000, 136-144.

Yoshinaga Yoshitaka. 'Nihon Rikugun heiki enkakushi: Murata ju kara rokketo made (5)' [A History of Japanese Army Weapons: (5) From the Murata Rifle to Rockets] *Gekkan JADI*, December 1995, 3-22.

INDEX

Military ranks are those at the end of career